Optical Materials

Optical
Materials

ROGER M. WOOD

THE INSTITUTE OF MATERIALS

Book 553
First published in 1993 by
The Institute of Materials
1 Carlton House Terrace
London SW1Y 5DB

British Library Cataloguing-in-Publication Data
Wood, Roger
Optical Materials
I. Title
620.11295

ISBN 0–901716–44–8

Typeset in Great Britain by
Fakenham Photosetting Ltd

Printed and bound in Great Britain at
The University Press, Cambridge

Contents

List of Symbols

a	absorption
A	Madelung's constant
$A\%$	absorption
A_{32}	spontaneous transition probability
B	absorption coefficient
c	velocity of light in air $= \lambda\nu = \nu n = 2.998 \times 10^8$ m s^{-1} $= (\mu_0\epsilon_0)^{-\frac{1}{2}}$
C	specific heat
d_{im}	non-linear susceptibility coefficient
$\frac{dn}{dx}$	dispersion coefficient – the change of refractive index with wavelength
$dT(V)$	retardation induced by applied electric field, V
D	Maxwell's displacement vector
D	material diffusivity coefficient
e	electronic charge $= 1.602 \times 10^{-19}$ C
E	electric field strength
E	photonic energy hν
E_a	electron affinity of negative ions
E_d	energy density to damage
E_D	laser pulse energy density
E_T	input energy for laser threshold
g_B	Brillouin gain coefficient
g_R	Raman gain coefficient
h	Planck's constant 0 6.624 $\times 10^{-34}$ J s^{-1}
i	angle of incidence
I	ionisation potential of positive ions
k	crystal momentum wave numbers
k	'imaginary' refractive index
k	Boltzmann's constant 0 1.380 $\times 10^{-23}$ J K^{-1}
K	thermal conductivity
K_e	bulk dielectric constant $= \epsilon/\epsilon_0$
K_m	bulk magnetic constant $= \mu/\mu_0$
L_c	critical interaction length
LIDT	laser-induced damage threshold
m	free electron mass $= 9.10 \times 10^{-28}$ g
m^*	effective electron mass
M	reduced mass
n	real refractive index
n_T	temperature-dependent refractive index
n_0	zero order power-independent refractive index
n_2	second order power-independent refractive index
p	induced dipole moment of an atomic bond

p	complex refractive index $= n + ik$; $p^2 = K_m K_e$
p_0	permanent dipole moment
P	polarisation
P_B	simulated Brillouin scattering (SBS) power threshold
P_C	beam power necessary to produce focusing at the Fresnel length
P_d	power density to damage
P_D	laser beam peak power density
P_R	simulated Raman scattering power threshold
r	angle of refraction
r	interionic distance
R_p	reflectance of p (parallel) polarised light
R_s	reflectance of s (perpendicular) polarised light
$R\%$	reflectance
t	time
t_d	diffusion time
T, T_0	phase retardation
T	temperature
T_A	ambient temperature
T_M	melting point
$T\%$	transmittance
v	velocity of light in a medium $= (\mu\epsilon)^{-\frac{1}{2}}$
V	electric field
W_0	minimum beam radius of focused beam
z	impedance of dielectric
z_{foc}	self-focusing length
z_0	impedance of free space

α	polarisability (or deformability of a molecule)
β	laser material gain coefficient
Δv	transition line width
ϵ	dielectric constant $= p^2 = n^2 - k^2$
ϵ_0	free space permitivity $= 8.854 \times 10^{-2} \text{ F m}^{-1}$
θ_B	Brewster angle
θ_c	critical angle
κ	thermal conductivity
λ	wavelength
λ_0	reference wavelength
μ	magnetic permeability
μ_0	magnetic permeability of free space
v	frequency of radiation
v_0	reference frequency
ρ	density
σ	conductivity
τ	pulse width
χ	susceptibility tensor

CHAPTER ONE

Introduction

This monograph is one of a series commissioned by the Institute of Materials. The series is aimed at providing a foundation for the better understanding of the science underlying today's technological society and as an introduction to each of the subjects in the series. It is hoped that not only will the reader find each of these monographs complete in themselves but that they will act as a spur for further study.

It has been written to act as the introduction to the subject of optical materials. It must be stressed that it is only an introduction and refers to a lot of work and further reading if the reader wants to go further in his or her understanding. Nevertheless it is hoped that it is complete enough to enable the interested reader to gain a better understanding of the topic.

When electromagnetic radiation impinges upon a material it interacts by reflection, transmission, absorption or scatter. The optical properties of materials are physical phenomena, in as far as the interaction of the electromagnetic radiation with the material is reversible, although they are firmly bound up with the chemical, atomic and crystallographic properties of the materials involved. The atomic, molecular and crystallographic arrangements also affect the other physical, thermomechanical and chemical properties of the materials and so it is not surprising that there are families of materials which exhibit particular properties, usually with a graduation of magnitude. For instance, most oxide materials are transparent in the visible and near infra-red and absorb above ~2 μm whilst fluorides, sulphides and nitrides have wider transparency windows and most materials from the II–VI and III–V regions of the periodic table exhibit semi-conducting properties. Cubic crystals exhibit linear optical characteristics whilst hexagonal and non-symmetric structures (and particularly chain structures) exhibit non-linear effects. The optical properties range from linear optical behaviour through a range of non-linear optical effects and on to electro-optic, opto-electronic and acousto-optic effects.

The dependence between the optical properties and the atomic, molecular and crystallographic structure is outlined in Chapter 2 and the resulting optical and physical properties are outlined in Chapter 3. This monograph will limit itself to those properties which are reversible and will stop once the energy absorbed is great enough to melt the material or crystal under

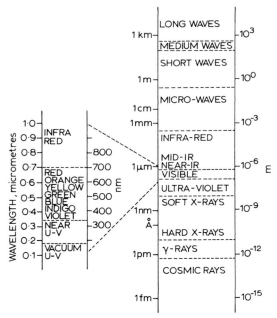

Figure 1.1   The electromagnetic spectrum

consideration, or the power density is enough to cause avalanche ionisation and electromagnetic breakdown in the material.

Although much of the theory is relevant to a wider selection of materials this monograph has specifically been limited to materials which are transmissive to radiation in the ultra-violet (u-v), visible and infra-red (i-r) regions of the electromagnetic spectrum, see Figure 1.1. An additional limitation is that it does not refer to 'glasses' in any detail as these have been the subject of a similar monograph (Rawson 1993). Therefore, although many optical materials are glassy and other materials which do not transmit in these regions still have optical properties (e.g. metals which absorb and reflect but do not transmit) no further specific mention will be made to these materials.

The treatment will include both single crystalline and polycrystalline, linear and non-linear, organic and inorganic materials. They include 'ordinary' (linear) optical materials (Chapter 4) but emphasis will be put on the great number of 'optically active' materials which are becoming increasingly available and which are leading the optical technology revolution which is in turn changing much of the commercial and military technology which is being introduced into the marketplace today. They include optically non-linear materials (Chapter 5), laser host crystals (Chapter 6), detector materials (Chapter 7), integrated optics and fibres (Chapter 8) and liquid crystals (Chapter 9). The monograph finally ends with a discussion of the power handling limits of optical materials (Chapter 10).

CHAPTER TWO

Molecular and Crystal Structure

2.1 Introduction

The basic building blocks which make up the universe as we know it are called elements and consist of positively charged nuclei (protons and neutrons) embedded in a negatively charged electron cloud. All these are in a state of perpetual animation, the motion of the electrons being much more rapid than that of the nuclei on account of the great difference between their respective masses. These atoms join together to form molecules and because of the dissimilarity of the nuclei have the potential to have associated dipole moments. The molecules in turn agglomerate themselves together, the most stable arrangement being that with the lowest potential energy. The resulting molecular arrangements are regular (crystals). However if there is not a seed crystal present materials will crystallise as randomly oriented crystallites and form polycrystalline materials. If single crystals are required then it is necessary to seed the cooling liquid phase with a single crystal. Even this does not always yield single crystals as some crystals exhibit phase changes (change of lattice parameter and/or molecular arrangements as they cool). In the case of perfect single crystals some of the molecular arrangements form with the dipole moments parallel, some are apparently random and some are anti-parallel. The two latter cases yield symmetrical properties (and consequently have linear optical properties) whilst materials with dipole moments formed up in parallel exhibit optical activity. For instance, NaCl (salt) crystallises in a regular cubic structure. Therefore even though the NaCl unit cell will have an asymmetry (and therefore an associated dipole moment) and NaCl crystal does not. Other materials (for instance, Al_2O_3 – sapphire) crystallise as a hexagonal arrangement in one plane (and as a linear lattice in the others) and are optically birefringent. Other materials crystallise so that they are not naturally birefringent but are distorted in the presence of an external optical or electric field. This class of materials is optically active and is termed electro-optic. When such a material interacts with an electromagnetic wave the molecule is deformed, the positive nuclei being attracted and the negative electrons repelled. Therefore, even when there is no permanent dipole moment, a dipole moment is induced.

2.2   Crystal symmetry

The molecular structure (both molecular size and arrangement) determines the resulting crystal structure and the resulting symmetry or asymmetry affects the physical properties. All possible crystal symmetries can be grouped into seven classification systems (see Figure 2.1) and these can be further subdivided into fourteen Bravais lattices. Using mathematical group theory, crystallographers further subdivide all crystal arrangements into 32 symmetry classes termed point groups. Figure 2.1 and Table 2.1 indicate how the point groups, Bravais lattices and symmetry systems relate to each other (Dekker 1960, Phillips 1962, Higgins 1992). All physical and chemical properties are determined by the interrelationship between the molecules and therefore the precise arrangement of the molecules in the space lattice affects the optical properties of the crystal including playing an important role in characterising

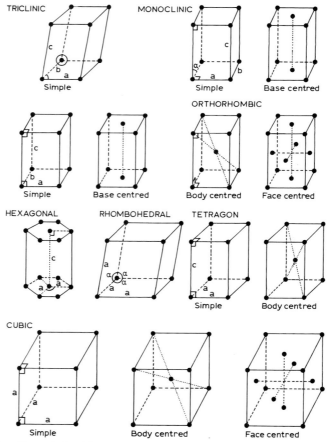

Figure 2.1   Crystal classification systems

Table 2.1   Symmetry properties of the seven crystal systems

Crystal system	Essential symmetry	Unit cell specification	Bravais lattice	Point groups
1. Triclinic	No planes, No axes	$a \neq b \neq c$ $\alpha \neq \beta \neq \gamma \neq 90°$	Triclinic, simple	$1, \bar{1}$
2. Monoclinic	One 2-fold axis *or* 1 plane	$a \neq b \neq c$ $\alpha = \beta = 90° \neq \gamma$	Monoclinic, simple Monoclinic, base centred	$2, m, 2/m$
3. Orthorhombic	3 mutually perpendicular 2-fold axes *or* 2 planes intersecting in a 2-fold axis	$a \neq b \neq c$ $\alpha = \beta = \gamma = 90°$	Orthorhombic, simple Orthorhombic, base centred Orthorhombic, body centred Orthorhombic, face centred	$222, mm2$ mmm
4. Rhombohedral (trigonal)	One 3-fold axis	$a = b = c$ $\alpha = \beta = \gamma \neq 90°$	Rhombohedral	$3, 32, 3m$ $\bar{3}, \bar{3}m$
5. Tetragonal	One 4-fold axis *or* a 4-fold inversion axis	$a = b \neq c$ $\alpha = \beta = \gamma = 90°$	Tetragonal, simple Tetragonal, body centred	$4, \bar{4}, 422$ $4mm, \bar{4}2m$ $4/m, 4/mmm$
6. Hexagonal	One 6-fold axis	3 coplaner axes at $120°$ $c \perp a \quad c \neq a$	Hexagonal	$6, \bar{6}, 622, 6mm$ $6m2, 6/m, 6/mmm$
7. Cubic	Four 3-fold axes	$a = b = c$ $\alpha = \beta = \gamma = 90°$	Cubic, simple Cubic, body centred Cubic, face centred	$432, \bar{4}3m, 23$ $m3, m3m$

the second- and third-order non-linear susceptibilities (see section 3.5). It is instructive to note that the second-order non-linear susceptibility vanishes for 11 of these 32 point groups due to centro-symmetry. In other words, the third rank susceptibility tensor reduces to zero for centro-symmetric crystals and second-order non-linear activity, such as SHG, is therefore impossible for these crystal classes. However, the fourth-rank susceptibility tensor does not reduce to zero for centro-symmetric crystals, indicating that third-order non-linear effects can occur in these materials. The proper evaluation of the crystal symmetry is therefore an important factor in determining the associated non-linear optical properties.

All crystals, except those belonging to the cubic system, are anisotropic to some degree and have optical and thermo-mechanical properties which vary depending on the direction. In the case of uniaxial crystals (such as quartz or calcite) there is a single direction termed the optic axis which is an axis of symmetry with respect to both the crystal structure and the orientation of the atoms within that structure.

2.3   Molecular Polarisability

The magnitude of the induced dipole moment, p, of an atomic bond increases linearly with the field strength E, and can be represented by the equation

$$p = \alpha E$$

where α is called the deformability or polarisability.

The total polarisation, P, per unit volume (containing N molecules) in an external field, E, is given by

$$P = \alpha N E$$

However, the polarisation, P, is also connected with Maxwell's displacement vector, D, by the relationship

$$D = E + 4\pi P \text{ and } D = \epsilon E$$

where ϵ is the dielectric constant and therefore, to a first approximation (in a gas)

$$\epsilon = 1 + 4\pi N\alpha$$

In a solid the presence of the surrounding molecules will have an effect on the displacement of the electron cloud around the nuclei. In general, molecules with symmetrical structure have no dipole moment while asymmetric molecules do possess one. In addition the arrangement of the molecules with relation to their nearest neighbours can also enhance or negate the dipole-moment effect. The following equation can be used to state the effect of the molecular structure/lattice and the ambient temperature, T.

$$\epsilon = 1 + 4\pi N(\alpha + p_0^2/3\,kT) = \epsilon_0 + 4\pi Np_0^2/3\,kT$$

where ϵ_0 is the dielectric constant for the case of a vanishing permanent dipole moment, $p_0 = 0$, and k is Boltzmann's constant,

$$\text{and } n - > (\epsilon_0)^{\frac{1}{2}} = (1 + 4\pi N\alpha)^{\frac{1}{2}}$$

where n is the refractive index of the material.

2.4 Dispersion

When an electromagnetic wave is incident on an atom or molecule the periodic electric force of the wave sets the bound charges into an oscillatory motion at the same frequency as the wave. The phase of this motion relative to that of the impressed electric force will depend on the impressed frequency and will vary with the difference between the impressed frequency and the natural frequency of the bound charges. When a light beam traverses the medium the amount of light scattered laterally is small because the scattered wavelets have their phases so arranged that there is practically complete destructive interference. On the other hand the secondary waves travelling in the same direction as the original beam do not cancel out, but combine to form sets of waves moving parallel to the original waves. The secondary waves must be added to the primary ones and the resulting amplitudes will depend on the phase difference between the two sets, thus modifying the phase of the primary waves equivalent to a change in their wave velocity. Since the wave velocity is the velocity at which a condition of equal phase is propagated, an alteration of the phase by interference changes the velocity of transmission of light through the medium. As the phase of the individual oscillators, and hence of the secondary waves, depends on the impressed frequency so the velocity of the radiation through the medium varies with the frequency of the probe light.

If it is assumed that the medium contains particles bound by elastic forces so that they are constrained to vibrate with a certain definite frequency v_0 then it is possible to postulate the shape of the dispersion (change of refractive index with frequency) curve. The passage of the light wave through the medium exerts a periodic force on the particles causing them to vibrate. If the frequency of the radiation, v, does not agree with v_0 then the vibrations will only be forced vibrations and will be of low amplitude. As the frequency of the radiation approaches v_0 the response of the particles will be greater and a very large amplitude vibration will occur when $v = v_0$. An equation relating the refractive index n, and the wavelength, λ, has been obtained (Sellmeier):

$$n^2 = \frac{1 + A\lambda^2}{\lambda^2 - \lambda_0^2}$$

In this expression the constant A is a material parameter and is proportional to the number of oscillators capable of vibrating at frequency v_0 and λ_0 is related to the natural frequency of the particles by the equation $v_0\lambda_0 = c$ where c is the velocity of light.

There is, of course, the possibility that the medium contains a number of different particles all having their own natural frequencies.

If the expression is expanded the following equation (Cauchy's) is gained

$$n = P + \frac{Q}{\lambda^2} + \frac{R}{\lambda^4} + \ldots$$

where P, Q and R are material constants.

Although Sellmeier's equation represents the dispersion curve very successfully in the transmission region of the medium it is not so successful at wavelengths where the medium has appreciable absorption. The difference between the Sellmeier curve and the experimentally measured curve has been shown to be due to the fact that it is necessary to take account of the energy absorption mechanism. This was done by Helmholtz who assumed that both absorption of energy and a frictional resistance to the vibration occurred. At the same time it should be remembered that a typical material contains a number of different molecules all having their own natural frequencies. Thus each molecule has its own absorption, a_i, at wavelength, λ_i, thus defining an absorption constant, k_0.

$$\text{Absorption constant} \qquad k_0 = a_i\lambda_i/4\pi$$
$$\text{Frictional force} \qquad g_i$$

$$n^2 - k_0^2 = 1 + \sum_i \frac{a_i\lambda^2}{(\lambda^2 - \lambda_i^2) + g_i\lambda^2/(\lambda^2 - \lambda_i^2)}$$

$$2nk_0 = \sum \frac{a_i/g_i\lambda^3}{(\lambda^2 - \lambda_i^2)^2 + g_i\lambda^2}$$

Although the curve of the refractive index against wavelength is different for every different material, the curves for all optically transparent media possess certain general features in common. This is illustrated in Figure 2.2 which purports to show the variation of the refractive index, n, from $\lambda = 0$ to several kilometers for a hypothetical substance.

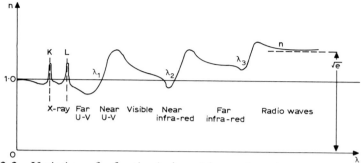

Figure 2.2   Variation of refractive index with wavelength

At $\lambda = 0$, $n = 1$.

For very short waves (gamma and hard X-rays) $n < 1$. An absorption is encountered in the X-ray region at a wavelength depending on the atomic weight of the heaviest element in the material. The refractive index rises sharply and then falls (the K-absorption limit of the element). After this lie other absorption discontinuities L, M, etc. as well as K, L, $M \ldots$ absorption limits of other elements present in the material. These absorptions are attributable to the innermost electrons in the atoms (the K, L, M, $\ldots$ shells) which are of decreasing energy (increasing distance from the nucleus). Because they are deep within the atom these electrons are shielded from the effects of collisions and electric fields due to neighbouring atoms and are hence sharp.

The refractive index continues to drop throughout the soft X-ray region until it reaches a broad absorption region in the ultra-violet, λ_1. This is due to the interaction of the light with the outer electrons of the atoms and molecules which make up the material. As these are not shielded the absorption band is broad. In the case of molecular gases the bands consist of a large number of sharp individual rotational bands which are collectively unresolved.

After this absorption peak the refractive index again drops (but is now >1). The nearer the ultra-violet absorption peak is to the visible the greater the dispersion $(\mathrm{d}n/\mathrm{d}\lambda)$ in this region.

In the near infra-red the refractive index again begins to drop more steeply until it runs into an absorption band centred at λ_2. This band is associated with the natural frequency of the lightest atoms in the material. They possess lower vibration frequencies than those associated with the electrons since the nuclei are appreciably heavier. Beyond this band there are usually others which are associated with even heavier molecules. The index of refraction increases each time these bands are passed and thus the index will be higher in the far infra-red than in the visible. At wavelengths beyond all the infra-red absorption bands the index decreases very slowly until it reaches a limiting value for very long wavelengths. The limiting value of the refractive index is $\sqrt{\epsilon}$, the ordinary dielectric constant of the medium.

It is possible to calculate the position of the absorption edges using the following arguments. On the short wavelength side transmission is restricted by electronic excitation and at long wavelengths by atomic vibrations and rotations. The width of the transparent spectral range increases as the energy for electronic excitation is increased and that for atomic vibration decreased (heavier atoms).

For ionic crystals the energy of electronic excitation is given by:

$$h\nu \sim \frac{Ae^2}{r} + E_a - I$$

where h is Planck's constant

$\nu = c / \lambda$ is the frequency
A is Madelung's constant
e is the electronic charge
r is the interionic distance
E_a is the electron affinity of negative ions
I is the ionisation potential of positive ions

This equation can be used to calculate the low wavelength absorption edge and shows that materials with strong bonding are transparent in the ultra-violet. Table 3.1 (p.18) lists the absorption edges of a number of commonly used materials in the ultra-violet. This absorption edge is affected by temperature as it is influenced both by the shift of the absorption peak (to longer wavelengths) and by broadening of the absorption band (Smakula 1962).

In covalent materials the position of the absorption edge is determined by the effective, m^*, and the free, m, electron masses and the refractive index, n:

$$\nu = \text{const.} \frac{1}{n^4} . \frac{m^*}{m}$$

Figure 2.3 shows the short wavelength absorption edge as a function of the refractive index for a number of covalent materials.

The infra-red absorption edge is set by vibration of the dipole moments of the atoms. The binding forces of the ions are of the same order of magnitude as those of the electrons, but as their masses are $^-10^4$ times greater the vibrational frequencies are about 10^2 times smaller.

$$\nu = \text{const.} / \sqrt{f/M}$$

where f is a force constant, M is the reduced mass, given by

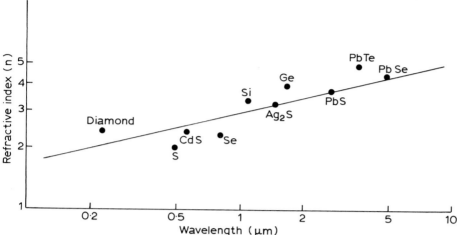

Figure 2.3   Short wavelength absorption edges and refractive indices of covalent elements and compounds

$$\frac{1}{M} = \frac{1}{M_1} + \frac{1}{M_2}$$

where M_1 and M_2 are the masses of ions with opposite charges.

These equations indicate that in order for materials to be transparent over a wide range of the infra-red they must have weak binding and large atomic mass. The infra-red absorption edge shifts towards longer wavelengths with increasing atomic mass (and also a decrease of the melting point). The intensity of the vibrational band is much weaker in partially ionic crystals than in pure ionic crystals and they are therefore transparent further into the infra-red. Pure covalent crystals of monatomic elements should, in theory, not exhibit any longwave absorption. The longwave absorptions are included in Table 3.1, indicating that there is either a certain amount of ionic bonding or lattice distortion in these crystals.

2.5   Effect of Irradiation

All crystals contain a number of lattice defects (vacancies, aggregates of vacancies, interstitials, dislocations, extrinsic impurities). The presence of these defects alter the charge distribution and therefore modifies the electronic levels in the vicinity of the defects (Dekker 1960). The simplest example of this is a positive ion vacancy in a cubic lattice (e.g. NaCl). The $3p$ electrons of the Cl⁻ ions neighbouring the vacancy will not be as strongly bound as normal because the positive ion vacancy acts as a negative charge. The outer electrons of the nearest six Cl^- ions form a potential well and therefore a positive hole in the electromagnetic energy level diagram. Trapped holes of this kind are called *V*-centres. Similarly, it is possible to form negative ion vacancies, termed *F*-centres, where an electron may be trapped. Even more complicated lattice defects will again modify the absorption spectrum. These absorption bands usually lie in the tail (long wavelength end) of the first fundamental absorption band. As this tail is usually in the ultra-violet or short wavelength visible region the new absorption is usually in the visible, giving rise to a colour cast in the crystal (e.g. LiF looks pink, KCl looks violet, NaCl looks brown-yellow). As the presence of a 'colour centre' denotes that there is either a positive or a negative ion vacancy it also denotes that the fundamental energy level diagram has been distorted and that the optical properties may be modified. This has been observed in both the laser host material Nd : YAG (see Chapter 6) and the electro-optic crystal lithium niobate. Both these materials have a well catalogued history of colour centre formation when they are irradiated with high energy, low wavelength light. However, although both crystals show this extraneous absorption on irradiation it can also be removed by thermal irradiation. It must be noted, however, that this annealing out of the colour centres must be done under controlled conditions as heating in the wrong atmosphere can induce even more complicated colour centres.

2.6   Polymers

Polymers were first developed as a by-product of the petrochemical industry for their flexibility and ease of processing, especially in thick film form. There has always been the conviction that it should be possible to chemically engineer materials with a great range of physical properties as it is possible to modify the structure of the polymer chains at will. Considerable work has gone into this and it has been shown that polymers can be engineered in such a way as to yield electro-optic, insulating, semi-conducting, and photo-luminescent properties (Meredith 1986, Nayar and Winter 1986, Donald and Windle 1992, Friend *et al.* 1992). For example, high purity conjugated polymers have been shown to exhibit strong photoluminescence and have been used as the 'active' layer in LEDs. Conjugated polymers contain repeat units made up of atoms with sp_2 and π covalent bonds such that there is an overlap of electron orbitals along the electron chain (Friend *et al.* 1992). These polymer structures have an alternation of single and double bonds along the chain and depending on the precise chemistry can be made to exhibit both semi-conducting and highly conducting characteristics. In general double and triple bonds provide the essential features required to give semi-conducting properties in carbon-based molecules. Some of the applications will be expanded upon in more detail in Chapters 5 and 6. It must be emphasised that the chemical engineering route is not a trivial exercise as unless the polymer is very pure the presence of other species affect the physical properties markedly.

One of the principal advantages of polymer engineering is that polymers can be easily fabricated in thick film form while in the liquid phase and then hardened. Nevertheless both single crystal and bulk forms have been used to provide specific optical characteristics. In many applications it is necessary to ensure that the polymer chains lie precisely parallel to each other. Niche applications have been found in large area luminescent screens and displays and in non-linear crystals and waveguides.

CHAPTER THREE
Physical Properties

This chapter will attempt to lay a theoretical basis to the optical properties of materials. The general reader may well prefer to skip this at this stage and to return to it later once they have read further into the subject and have understood the necessity for this basis.

3.1   Electromagnetic Theory

All electromagnetic phenomena are governed by Maxwell's equations. This is not the time and place to derive these (see Jenkins and White 1957, Ditchburn 1958, Born and Wolf 1975, Ghatak and Thyagarajan 1989, Olver 1992 for good introductions to the subject), but it is necessary to stress that it is not possible to progress beyond the confines of this book without having at least a smattering of understanding and application of these equations. This section will confine itself to the application of a series of equations which have their derivation in the fundamental equations.

The statement that light is an electromagnetic wave is taken to be fundamental and has the implication that it is associated with transverse undulations of the associated electric (E) and magnetic (H) fields. These fields are subject to the laws of conservation of energy and this in turn allows us to predict and quantify the interaction of light with both bulk materials and the interfaces between materials.

v, the velocity of light within a medium, is related to the velocity of light in vacuo, c, by the reflective index, n.

$$c = v.n$$

This in turn is related through the bulk dielectric, K_e, and magnetic, K_m, constants to the dielectric permittivity, ϵ, magnetic permeability, μ, and conductivity, σ, Of the medium such that

$$c = (\mu_0\epsilon_0)^{-\frac{1}{2}}$$

$$v = (\mu.\epsilon)^{-\frac{1}{2}}$$

15

Where $K_e = \epsilon/\epsilon_0$
$\quad\quad K_m = \mu/\mu_0$
$\quad\quad p^2 = K_m K_e$
and p is the complex refractive index $= n + ik$

The wave impedance of the medium μ/ϵ $(=z^2)$ is a constant for the medium and is related to the oscillating electric, E, and magnetic, H, fields by the equations

$$E_x/H_y = z_0$$

and

$$E_x H_x + E_y H_y = 0, \ E_z H_z = 0$$

The solutions for the electric and magnetic components are of the form:

$$E_x = A e^{i\omega(t + pz/c)}$$

where A is a constant.

By substitution we find

$$E_x \sim e^{i\omega(t - nz/c)} \cdot e^{-\omega kz/c}$$

The first term is a sine wave and the second is an expotential decay damping. By further substitution we find

$$p^2 = \mu\epsilon c^2 - (\mu\sigma c^2/\omega) = K_m K_e - i(K_m\sigma/\omega\epsilon_0)$$
$$= n^2 - 2ink$$

The complex wave impedance of the medium, z_0, can be split into resistive and reactive components:

$$z_0 = E_x/H_y = \mu c/p = \mu c(n + ik)/(n^2 + k^2)$$
$$= \mu cn/(n^2 + k^2) + i\mu ck/(n^2 + k^2)$$

with a phase angle of $\tan\theta = k/n$ between E_x and H_y

$K_m = \mu/\mu_0 \approx 1$ for all insulators and therefore the dielectric constant $K_e \approx n^2$
The dielectric constant varies with frequency and results in dispersion of the electromagnetic wave.

In the case of a dielectric $\sigma = 0$, $\therefore k = 0$ and $n^2 = K_m K_e$

In the case of a metal $\sigma \gg 1$ and $n^2 \approx \frac{1}{2}K_m$. $K_e \sigma/\epsilon\omega \sim k^2$

3.2   Refractive Indices

The last section showed that the interaction between an electromagnetic wave and a bulk material is governed by the complex refractive index of the material in question. This in turn can be split into resistive and reactive components and the electric and magnetic fields can be seen to have two components, a regular sine wave with a modified period and a damping term with an exponential decay. The former term is responsible for the reflection characteristics of the medium while the latter term is responsible for the absorption. The precise values of the refractive indices are bound up in the physical and chemical structure of the medium and this was discussed in Chapter 2.

The refractive indices of a material, being bound up in the structure of the material, are not constant but change with wavelength of the probe light. A typical plot of refractive index versus wavelength is shown for zinc selenide in Figure 3.1. In this plot the transmissive region where $p = n$ is bounded at each extremity by an absorbing region, where $p = n + ik$. As the rest of the monograph is concerned with the interaction of the medium with the light

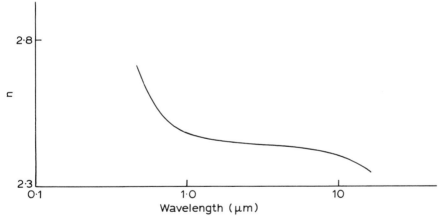

Figure 3.1   Refractive index versus wavelength: zinc selenide

wave in the transmissive region we are therefore mainly concerned with real values of the refractive indices and the complex part is taken to be small. In this regime $\sigma = 0$, $k = 0$, $p \approx n$ and $n^2 = K_m K_e$.

A list of the refractive indices of commonly used materials is shown in Table 3.1.

In the case of materials where the structure is not isotropic (ie the molecular spacing is different in different planes) the electromagnetic interaction will be different depending on the direction of the wave vector and the refractive indices will be different in the various directions. The transmission characteristics of the material will then change depending on the direction of the wave vector and the material will be said to be birefringent and the light beam will become polarised (see section 3.4). At high electric fields most materials

Table 3.1 Physical parameters of commonly used optical materials

Material	Chemical formula	Refractive index (n)	Wavelength $\lambda_{1\mu m}$	Wavelength $\lambda_{2\mu m}$	n_2 m^2V^{-2}	ϵ	Reflectance $R\%$	MP°C	Hardness knoop	Density gm cm^{-3}	Solubility gm per 100 gm
Barium fluoride	BaF$_2$	1.47	0.3	9.0	$1.12 > 10^{-22}$	7.32	3.6	1280	65–82	4.89	0.12
Calcium fluoride	CaF$_2$	1.45	0.3	9.5	$7 > 10^{-23}$	6.81	3.4	1360	120–200	3.18	0.0016
Magnesium fluoride	MgF$_2$	1.38	0.2	6.0	–	–	2.5	1498	415	3.18	0.008
Sodium fluoride	NaF	1.3	0.12	10.0	–	–	1.6	1270	60	2.79	4.2
Sodium Chloride	NaCl	1.54	0.17	14.5	$7.18 > 10^{-22}$	5.9	4.5	801	15	2.17	hygroscopic
Potassium chloride	KCl	1.45	0.18	23.0	$3.68 > 10^{-22}$	4.8	3.4	776	7.2	1.99	hygroscopic
Potassium bromide	KBr	1.53	0.3	25.0	–	–	4.4	1003	7	2.75	65.2
Quartz	SiO$_2$	1.45	0.25	1.25	$1.5 > 10^{-22}$	3.7	3.4	1600	461	2.21	insoluble
Sapphire	Al$_2$O$_3$	1.75	0.3	4.0	$1.44 > 10^{-22}$	10.15	7.3	2015	1705	3.986	$9.8 > 10^{-5}$
Garnet	Y$_3$Al$_5$O$_2$	1.82	0.4	1.3	$3.51 > 10^{-22}$	–	20.3	1970	1215	4.55	insoluble
Zinc sulphide	ZnS	2.26	0.3	11.0	–	8.37	15.0	1830	355	4.09	insoluble
Zinc selenide	ZnSe	2.45	0.5	16.0	–	9.1	17.6	1520	100–150	5.27	insoluble
Silicon	Si	3.4	1.3	6.0	–	11.7	30.0	1410	1150	2.33	insoluble
Germanium	Ge	4.0	1.8	16.0	–	16.0	36.0	938	190	5.32	insoluble
Gallium arsenide	GaAs	3.3	1.0	20.0	–	12.8	28.3	1238	750	5.31	insoluble
Cadmium telluride	CdTe	2.70	1.0	2.0	–	10.6	21.1	1090	45	5.85	insoluble

exhibit optical non-linearity. These are important facets of modern optics and will be further expanded in section 3.5 and in Chapter 5.

3.3 Reflection, Transmission and Absorption

The equations for the electromagnetic waves are defined as

$$E_x = e^{i\omega(t-nz/c)} \cdot e^{-i\omega z/c}$$

saying that the wave is transmitted through the material with an associatedelectromagnetic field of period nz/c and with an exponential absorption of $e^{-i\omega z/c}$.

When the wave falls at normal incidence on a surface between two media (see Figure 3.2) the boundary conditions are:

The normal components of the magnetic induction vector and the displacement vector are continuous and

The tangential components of the electric and magnetic field vectors are also continuous, i.e.

$$E_y + E''_y = E'_y, H_x + H''_x = H'_x$$

where E, H; E', H', E'', H'' are the electric and magnetic field vectors appropriate to the incident, transmitted and reflected waves (as defined in Figure 3.2) at the surface between the two media (with refractive indices n and n' respectively). But as

$$E_y/H_x = z, E'_y/H'_x = z', E''_y/H''_x = -z$$

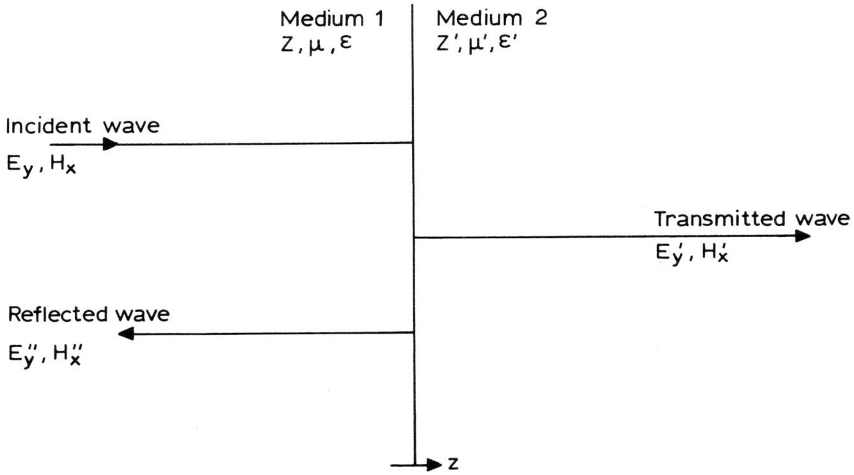

Figure 3.2   Interaction between electromagnetic waves at a surface

$$\therefore E_y''/E_y = (z' - z)/(z' + z)$$

$$E_y''/E_y = (n - n')/(n + n')$$

and
$$E_y'/E_y = 2n/(n + n')$$

The energy propagated in the wave (Poynting's vector) $= (\epsilon/\mu)^{\frac{1}{2}}$

$$= n.E^2$$

The proportion of the energy reflected is therefore (Fresnel's Law)

$$R = nE_y''^2/nE_y^2 = (n - n')^2/(n + n')^2$$

The proportion of the energy transmitted is

$$T = n'E_y'^2/nE_y^2 = 4nn'/(n + n')^2$$

and $R + T = 1$ as energy is conserved.

Note that both R and T are symmetrical and therefore the proportion transmitted or reflected is constant and does not depend on the direction of the incoming wave, i.e

if $n' > n$ then E_y''/E_y is $-ve$ the phase of the reflected wave is exactly out of phase with the incident wave;

if $n' < n$ then E_y''/E_y is $+ve$ both the incident and the reflected wave have the same phase.

A plot of the reflectance at normal incidence in air at the surface of a medium of refractive index n, for a range of values of n, is shown in Figure 3.3.

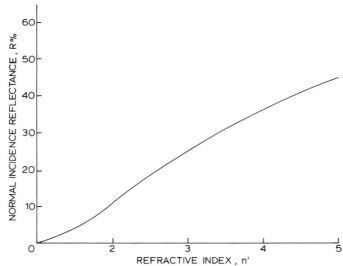

Figure 3.3   Normal incidence reflectance, $R\%$ for an air $(n = 1)$/material (n^1) interface as a function of the material refractive index n^1

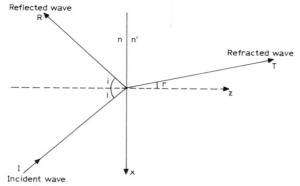

Figure 3.4   Ray diagram at an interface

Extension of the same argument leads to the derivation of all the laws of physical optics (using the terminology of Figure 3.4).

e.g Snell's Law : $n\sin i = n'\sin r$

This equation defines the relationship between the angle of incidence, i, and the angle of refraction, r, as a ray passes from a medium with refractive index n into one with a refractive index n'.

When the incident ray is at non-normal incidence Fresnel's law has to be modified to take account of the angle of the material plane to the two plane polarised components of the incident unpolarised light. Although this derivation is outside the remit of this monograph (see Jenkins and White 1957) it is relevant to quote the equations connecting the intensities of the incident, reflected and refracted beams to the angle of incidence since these values ultimately depend on the refractive indices of the media. The Fresnel laws may be written as:

$$R_s = \frac{\sin^2(i - r)}{\sin^2(i + r)} \qquad R_p = \frac{\tan^2(i - r)}{\tan^2(i + r)}$$

and

$$R_p = \frac{4\sin^2 r\cos^2 i}{\sin^2(i + r)} \qquad T_p = \frac{4\sin^2 r\cos^2 i}{\sin^2(i + r)\cos^2(i - r)}$$

where subscript s refers to vibrations perpendicular to the plane of incidence and subscript p refers to vibrations parallel to the plane of incidence.

Examples of the reflectance at (a) an air/glass ($n = 1/n = 1.5$) interface and at (b) an air germanium ($n = 1/n = 4$) interface are shown in Figures 3.5a and b. It will be seen from these figures that the p-polarised ray goes through a minimum reflectance before rising to 100%R at grazing incidence

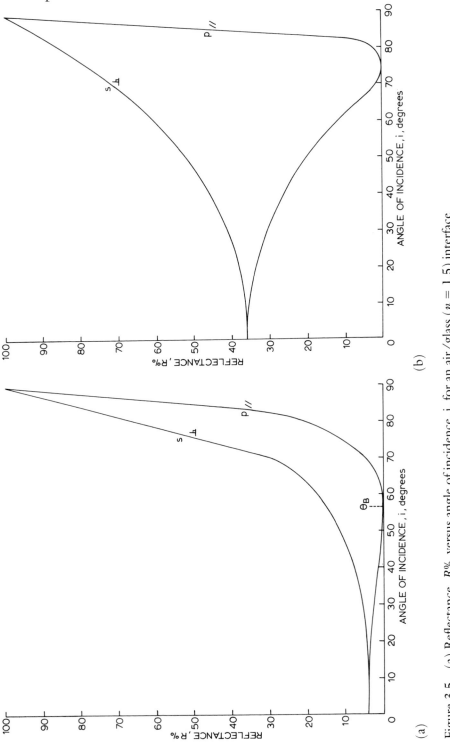

(a)

(b)

Figure 3.5 (a) Reflectance, $R\%$, versus angle of incidence, i, for an air/glass ($n = 1.5$) interface
(b) Reflectance, $R\%$, versus angle of incidence, i, for an air/germanium ($n = 4$) interface

while the *s*-polarised ray steadily increases from its normal incidence value to the 100%*R* when *i* = 90°. The angle of incidence where the *p*-ray reflectance reaches zero is called Brewster's angle and is widely used for producing polarised light.

$$\text{Brewster's Law: } \theta_B = i = \tan^{-1}(n'/n)$$

This is illustrated in Figures 3.5a and b and also in Figure 3.6 where the angle is plotted as a function of n'/n.

In the case of dense-to-rare, internal, reflection a critical angle, θ_c, exists which is defined as the angle where all the light is reflected and none transmitted.

$$\theta_c = \sin^{-1}(n/n')$$

This is also illustrated as a function of n'/n in Figure 3.6 and for the cases of glass/air and germanium/air as a function of angle incidence in Figure 3.7.

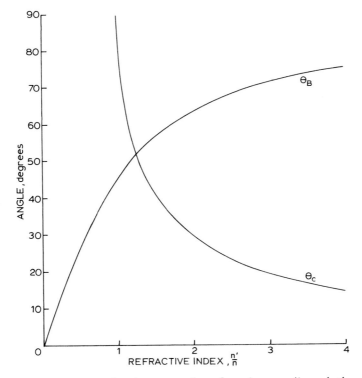

Figure 3.6 Brewster angle, θ_B, at an interface (*n* to *n'*) and the internal reflectance critical angle, θ_c, at an interface (*n'* to *n*) as a function of the material refractive index ratio n'/n

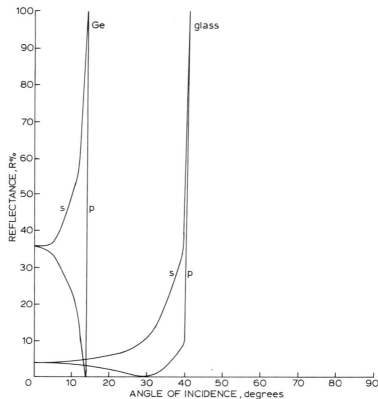

Figure 3.7   Internal reflectance, *R*%, versus angle of incidence, *i*, for glass-to-air ($n = 1.5 \rightarrow 1$) and germanium-to-air ($n = 4 \rightarrow 1$) showing the variations with plane of polarisation

3.4   Polarisation

Electromagnetic theory requires that the light wave vibrations are transverse, being confined to the plane of the wave front, whereas sound waves are longitudinal. When a light wave enters a medium which is asymmetric and has different refractive indices for different propagation directions the light resolves itself into two components and, in the general case, there will be two refracted beams instead of the single one observed in isotropic materials, these are termed the Ordinary (O) ray and the Extraordinary (E) ray. This is termed double refraction (see Figure 3.8). Rotating the crystal about the O-ray will cause the E-ray to rotate round this direction. The refractive indices are usually termed n_o and n_e for the two directions respectively.

The use to which birefringence can be put is most easily illustrated by explaining the design of a polarising prism. Such a prism is often required to

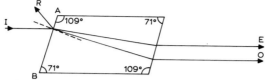

Figure 3.8   Double refraction of light in a birefringent material

ensure that the light beam is polarised in a known direction and can also be used as part of the switching arrangement for a high power pulsed laser (Q-switching). Two designs of polariser are shown in Figure 3.9 (a,b). (a) is a conventional Glan-air prism and (b) is a Brewster-faced Glan-air prism. In both designs the internal interface is cut so that the angle of incidence is greater than the critical angle for the ordinary ray, θ_{co} and less than critical for the extraordinary ray, θ_{ce}, (preferably the Brewster angle for the extraordinary ray θ_{Be}) so that, in theory, the E-ray is totally transmitted and the O-ray is totally reflected. The Brewster-faced design has two advantages – the first that the rays enter at the Brewster angle (and hence with no reflectance loss) and the second is that the rays are angularly separated by the time they reach the internal interface.

For calcite $\theta_{ce} = 42.5°$, $\theta_{co} = 37°$, $\theta_{Be} = 34°$. It will therefore be realised that, assuming that the input faces can be perfectly anti-reflectance coated, there will be an E-ray loss at the internal interface in the case of the Glan-air prism as $\theta_{co} > \theta_{Be}$ but not in the case of the Brewster-faced prism.

In order to allow the construction of a perfect Glan-air prism the polarising material must have refractive indices such that the Brewster angle for the E-ray is equal to or larger than the critical angle for the O-ray. This can be achieved

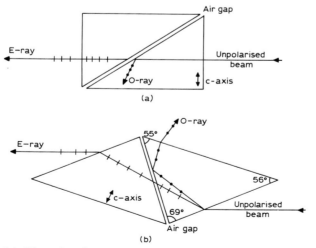

Figure 3.9   (a) Glan-air prism
(b) Brewster-faced glan-air prism

either by using calcite in the Brewster-faced design or by using a crystal with higher birefringence such that:

$$\theta_{Be} = \cot^{-} n_e > \theta_{co} = \sin^{-1}(1/n_o)$$

$$\text{or } n_o{}^2 - 1 > n_e{}^2$$

$$\text{or } \Delta n > \tfrac{1}{2}(n_o + n_e)$$

A graph showing the solution of this expression for equality is shown as Figure 3.10. The top line in this figure is the difference between refractive indices, $\Delta n = n_e - n_o$, necessary to fulfil the condition. Materials lying on or above the θ_{Be}

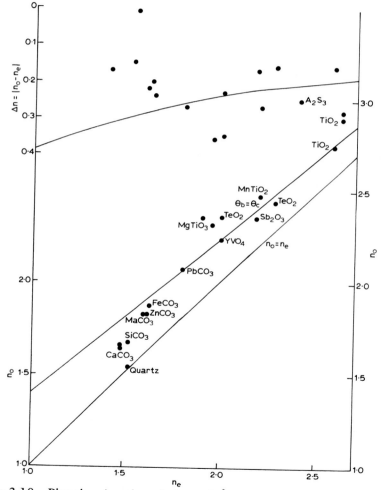

Figure 3.10 Plot showing the solution of $n_0^2 - 1 = n_e$ and the birefringence, Δn, necessary for $\theta_b = \theta_c$

$= \theta_{co}$ line, in principle, permit a virtually perfect polariser to be constructed. All birefringent materials with negative birefringence lie above the $n_e = n_o$ line; the more highly birefringent the crystal the higher it will lie on the plot. The required value of Δn increases as n increases. Since there is a general trend for both Δn and n to decrease with increasing wavelength it becomes more difficult to find materials to meet the $\theta_{Be} > \theta_{co}$ criterion at longer wavelengths. A list of the most highly birefringent materials is given in Table 4.2 (p.39) and these materials are included in Figure 3.10 along with the two most commonly used materials – quartz and calcite. The main reason that the latter two crystals are still used in the majority of polarisers is that the large single crystals necessary are readily available whilst those with higher birefringents are not.

When the light enters the crystal along the optic axis the double refraction effect disappears. When the light enters the crystal at right angles to the axis the E-wave, having vibrations parallel to the optic axis, will travel faster than the O-wave, but along the same optical path. The resulting phase difference is a measure of the distance the E-waves advance on the O-waves in a given thickness of crystal and is given by

$$\delta = 2\pi d(n_o - n_e)/\lambda$$

where d is the physical thickness of the plate.

This is widely used in the production of phase plates which are widely used for contrast enhancement in optical microscopy.

3.5 Optical Non-Linearity

The intrinsic transmission spectrum of a material is bounded at short wavelengths by multi-photon absorption and at longer wavelengths by band-gap absorption. Between these limits Rayleigh scattering and Fresnel reflection define the minimum transmittance at low optical input. It is not possible to increase the transmitted power indefinitely because of the advent of non-linear effects and ultimately laser induced damage.

As was explained in Chapter 2, materials can be thought of as a collection of charged particles: ion cores surrounded by electrons. In dielectric materials these particles are bound together – the bonds having a degree of elasticity – by the interaction between the charged particles. When an electric field is applied across the material the positive charges (ions) are attracted and the negative charges (electrons) are repelled slightly. This results in the formation of a collection of induced dipole moments and hence induced polarisation. The electric field can be in the form of a voltage across the material or as a light beam passing through the material or as a combination of both.

A light wave comprises electric and magnetic fields which vary sinusoidally at 'optical' frequencies ($\sim 10^{13} \rightarrow 10^{17}$ Hz). The motion of the charged particles in a dielectric medium when a light beam is passed through it is therefore oscillatory and they form oscillating dipoles. As the effect of the electric field is much greater than the effect of the magnetic field the latter can, to all intents and purposes, be ignored. Also as the electrons have a lower mass than the ions it is the motion of the former which is significant at the high optical frequencies (ultra-violet and visible regions of the spectrum). The electric dipoles oscillate at the same frequency as the incident electric field and modify the way that the wave propagates through the medium.

The electric displacement is: $D = \epsilon_o E + P$, where ϵ_o is the free space permittivity and the polarisation, P, can be expressed in terms of the electric field, E, and the susceptibility, χ.

The dielectric constant $E = 1 + \chi$ and the refractive index $n = R_e / (1 + \chi)$

$$P = \epsilon_o \left(\chi^{(1)} E + \chi^{(2)} E^2 + \chi^{(3)} E^3 + \dots \right)$$

where $\chi^{(1)}$ is the linear electric susceptibility
$\chi^{(2)}$ is the quadratic non-linear susceptibility
$\chi^{(3)}$ is the cubic non-linear susceptibility

and the field dependent refractive index can be written as:

$$n^2 = 1 + \chi^{(1)} + \chi^{(2)} + \chi^{(3)} + \dots$$

The oscillation of the electromagnetic wave will result in an oscillation of the induced dipole moment which in turn will result in the generation of light having different frequencies to that of the incoming light wave. It should be noted that the applied electric field used in the above equation is the total field and can be made up of a number of contributions (i.e. both optical and electric). The main application of the non-linear effects along with their relevant susceptibility term are further discussed in Chapter 5.

It should be noted that the motion of the charged electrons in the dielectric medium can only be considered to be linear with the applied field if the displacement of the electrons is small. When the incident field is large in comparison with the internal field which binds the electrons and the ions together (typically $E \sim 3 \times 10^{10}$ Vm^{-1}) significant non-linearity of the P/E curve will occur. This is synonymous with optical intensities of $\sim 10^{14}$Wcm^{-2}. Intensities of this magnitude can be reached by focusing high power short pulsed laser beams into a material.

The susceptibility tensor, χ, can be expressed as a matrix made up of the relevant non-linear susceptibilities, d_{im}. These are discussed more fully in Chapter 5.

3.6 Self-focusing

Self-focusing is a reduction of the laser beam diameter below the value predicted from the refractive index of the unirradiated material and the laser beam/optics. It can be caused by any process which leads to an increase in the index of refraction, such as increasing light intensity or temperature.

The refractive index of a material can be expressed as:

$$n = n_0 + n_T(T) + n_2 E^2 +$$

where n_0 = zero-order power independent *ri*
n_T = temperature dependent *ri*
n_2 = second-order power dependent *ri*

There are a series of self-focusing mechanisms including electrostriction, electronic distortion, molecular distortion, molecular libration and absorption heating. These operate under different time relaxation regimes and the process is a complicated amalgam of power density and pulse length dependence. In general self-focusing only occurs at extremely high optical power densities and can lead to laser-induced damage (see Chapter 10). This filamentary damage threshold (David 1970) is related to a self-focusing length which is a function of the non-linear refractive index n_2. When the sample under test is longer than this self-focusing length the filamentary damage threshold is a minimum.

In the thermal case (continuous rays, cw, and the long pulse regimes) the self-focusing length, Z_{foc}, can be calculated as the point where the hot (slower) ray travelling along the beam axis meets in phase with the cold (faster) ray arriving from the edge of the beam (radius r).

$$Z_{foc} = \frac{r^2 n \pi}{\sqrt{2 n_T P_D \tau}}$$

where P_D is the power density of the beam
and τ is the pulse length

$$\Delta n = n_T P_D \tau = \frac{n P_D \tau A}{\rho C}$$

where ρ = density
C = specific heat
A = absorption coefficient

As Fresnel diffraction also takes place, trapping of the beam is defined as occurring when the self-focusing distance, Z_{foc}, is less than or equal to the Fresnel length (where diffraction doubles the beam diameter). The power in the beam necessary to produce focusing at the Fresnel length is P_c, where

$$E_p = P_c \tau = \frac{(1.22\lambda)^2 n \pi}{32 n_T}$$

where n_T is the energy dependent *ri* factor.

It should be noticed that in this case the pulse energy, Ep, required to trap the beam is a constant independent of pulse length.

Electrostriction occurs in a dielectric material under laser irradiation as the net electrostrictive force at any point is proportional to the square of the electric field. Thus a radially symmetric beam would lead to a radially symmetric stress with an associated change in the refractive index leading to self-focusing. As the acousto-optical interaction involves a radially propagating compression wave driven by the light intensity, beam focusing is caused by the compressional increase in refractive index along the beam axis. Electrostrictive self-focusing can therefore occur even in non-absorbing materials. The time taken for the acoustic response to develop is of the order of 10^{-10}s (Bliss 1971).

For free lasing and Q-switched pulses the threshold intensity is independent of the pulse length, the critical threshold power being given by:

$$P_c = \frac{C.n_0\lambda.A}{3.4\pi^2.n_2.L_c}$$

where L_c = critical interaction length

For a single mode-locked pulse the power density necessary for damage increases with decreasing pulse length. A mode locked train, however, gives the acoustic response time to develop and the threshold therefore occurs at similar power densities to that in the case of the longer pulses.

Other self-focusing processes, such as electronic distortion and molecular libration, have even shorter relaxation times and affect the refractive index even for picosecond pulses and consequently the damage thresholds are independent of pulse length.

The effects of self-focusing are obviously only evident when a light beam is passed through an optical material at high power and energy densities. These conditions are met in many laser systems, such as laser resonators, and in fibre-optic cables. The largest laser systems (e.g. Shiva, Nova and the Rutherford) are therefore built up starting with crystal oscillators, graduate to glass rod amplifiers and finish with face-pumped glass slab amplifiers of limited thickness. Similarly, even though optical fibres are only called upon to handle much lower total energies, the power densities involved, once the laser beam is focused into the fibre, can reach the self-focusing threshold. This along with the onset of Brillouin and Raman scattering (see Chapter 5) limits the power which can be transmitted down an optical fibre. This has widespread implications in the military, medical, communication and material processing fields. Useful articles have been published on this subject by Zverev *et al.* (1969, 1970), Bliss (1971) and Soileau *et al.* (1980, 1989).

CHAPTER FOUR

Optical Properties of Linear Materials

4.1 Introduction

This chapter outlines the optical properties of linear materials (ones that do not change with input power density). It will mainly be a list of the properties, building on the theory developed in Chapters 2 and 3. The materials will be grouped into similar classes and it is hoped that the reader will be able to perceive the sometimes subtle relationship between the different atoms and the measured optical properties. Spectral transmission plots and lists of the most common physical parameters are given for the most common optical materials. Fuller tables of properties may be found in Moses (1971), Donaldson and Edwards (1983), Savage (1984), Wolfe *et al.* (1985) and in the Chemical Rubber Company Tables (1989).

4.2 Halides

After the silicate glasses these form the largest group of visible transmission optical materials. A list of the most pertinent optical and physical properties of the more common halides was included in Table 3.1. Halides mostly have a short wavelength ultra-violet absorption edge, a lengthy transmission region (commensurate with a large atomic energy gap) and a vibrational absorption in the far infra-red. Most halides crystallise in a regular cubic structure and therefore exhibit linear optical properties. With these advantages it is clear that if their other physical properties were usable then they would be even more widely used than they are today. Unfortunately they are mainly soft or easily cleaved crystals, hygroscopic and suffer from the formation of optically induced absorption (colour centres). These characteristics limit their usefulness, particularly when their environment cannot be closely controlled. They include the following:

Barium fluoride (BaF_2).This crystal is widely used in the visible and infra-red regions of the spectrum.

Calcium fluoride (CaF_2). This material is of limited use as it has only a

31

restricted transmission range although it is much less soluble than the equivalent barium salt.

Magnesium fluoride (MgF_2). This material is one of the most widely used coating materials as it can be readily evaporated, using a thermal source, and forms a dense, semi-hard and relatively impervious layer. It is, in particular, a nearly perfect match for Nd doped YAG as the refractive index $1.38 = \sqrt{1.90}$ (where $n = 1.90$ is the refractive index of Nd:YAG) and therefore very nearly acts as a perfect anti-reflection for this crystal.

Sodium chloride (NaCl). This is one of the most naturally abundant crystals but suffers badly from cleavage planes, is easily solarised and is very hygroscopic.

Potassium chloride (KCl). This is perhaps the most widely used halide as it has many spectroscopic applications. It crystallises easily into large clear boules and has an extremely wide transparency range. It is however, hygroscopic and is best used inside sealed units.

Lead fluoride (PbF_2). This material, in conjunction with the very similar lead chloride, is one of the most promising of the halides in that not only does it have a useful transmission range (to beyond $10\,\mu m$) but it is also less hygroscopic than the rest of the halides. There is some residual birefringence and this is best avoided by the use of hot pressed polycrystalline rather than single crystal material.

4.3   Oxides

These materials are the most used in the visible region. They include the common glasses, quartz, sapphire, garnets and vanadates. In general they are transmissive in the visible region and most start to absorb heavily in the near infra-red. They are all moderately hard, insoluble and cleanable and can be used in the open atmosphere. They include the following:

Quartz (SiO_2). This crystallises in a hexagonal structure and therefore has an optic axis and is birefringent.

Sapphire (Al_2O_3). This also crystallises in an hexagonal structure and is birefringent. It is hard, transparent into the infra-red and is insoluble. It is widely used as a substrate because of its superior optical and thermal properties for applications where the cheaper alternatives fail. It can be doped with the whole range of the Group II transition elements and forms the basis for the main range of precious stones.

Garnets (e.g. yttrium aluminium oxide, YAG). These materials are generally very hard (undoped YAG is commonly called triagem or artificial diamond as it is very hard, clear and has a relatively high refractive index).

Vanadates (e.g. chromium vanadate). These materials form the main range of phosphors.

4.4 Chalcogenides

These materials, which include the widely used zinc selenide, zinc sulphide, germanium and silicon, have been developed to a very high standard particularly because they offer high transmission in the infra-red (Savage 1984, 1985). However, they are only moderately hard (easily scratched), some are soluble in water and they all tend to suffer from the phenomenon of thermal runaway (Wood 1988).

Very large, high quality boules of germanium and silicon are available as these have been developed for large-scale semi-conductor circuitry. However, it must be stressed that this latter application does not necessarily lead to high optical transmission material as the doping requirements are different. All these materials attract a surface layer of water which modifies the absorption properties. This water can be either heated off or can be removed by cleaning with a solvent such as iso-propyl alcohol.

Zinc sulphide and zinc selenide have both been developed to a high optical standard since they are widely used for the fabrication of thermal imager lenses and domes. For many years it was thought that they were intrinsically yellow in colour and it was not until it was proved that this absorption could be removed in about 1980 that they became firmly established. They are usually prepared as polycrystalline material by large-scale chemical vapour deposition (Fradin and Bua 1974). The result of this preparation technology is that the material is not completely isotropic and is formed as hexagonal pillars. This is not necessarily bad except that if there are impurities present in the deposited material then these congregate at the hexagonal interfaces and can lead to the material having totally different absorption coefficients in different crystallographic directions.

Other chalcogenide materials include the tellurides and arsenides. Whilst being useful for optical designs because of their different refractive indices they do not offer any other specifically better optical or physical properties than the first four in the series.

4.4.1 Optical Properties of Germanium

Germanium is widely used as a window material in infra-red systems because of its low dispersion and absorption at room temperature in the wavelength region 2 to 12 μm. Below 2 μm the free carrier absorption increases rapidly as electrons are excited across the 0.785 eV band gap. The 12 μm upper transmission limit marks the onset of absorption by the crystal lattice. This is characterised by an extremely complex series of absorption peaks. The infrared absorption in the transparent region is a function of the free-carrier concentration (see Figure 4.1) (Wood *et al.* 1984). The small band gap gives

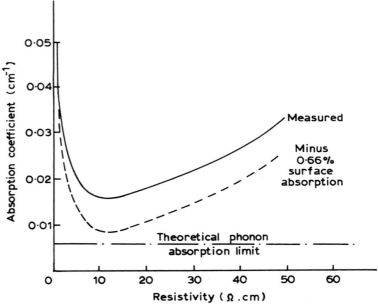

Figure 4.1   Variation of absorption with resistivity: germanium

rise to a relatively large population of intrinsic carriers at room temperature and causes irradiated samples to exhibit thermal runaway at temperatures in excess of about 50°C (see Figure 4.2). All semi-conducting materials exhibit this free-carrier absorption and associated thermal runaway but as they

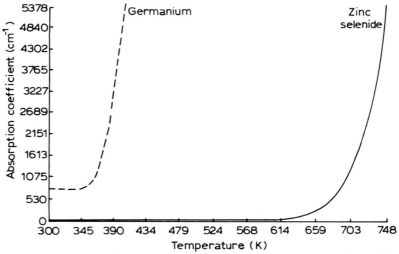

Figure 4.2   Variation of absorption with ambient temperature: Ge and ZnSe

have differences in band gap, thermal conductivity etc. the temperature at which thermal runaway is self-generating is also different from material to material.

4.5   Carbides, Nitrides and Phosphides

These materials have gradually become more available as the subject of durability has become increasingly more important both in the military and the commercial spheres. In all cases the primary impetus for their development has been as durable thin film materials but it can reliably be forecast that as long as the fabrication technology can be overcome then one or more of these materials will supersede many if not all the preceding materials. In general the materials are transparent through much of the visible layer and contain a fundamental absorption peak in the infra-red. This can be traced directly to the anions and occurs at longer wavelengths the heavier the anion.

4.5.1.   Diamond

Single crystal diamond occurs naturally as a face-centred cubic crystal. As such it is isomorphous and can be used as an isotropic material. Natural type IIA diamond has probably the widest transmission spectrum range of any material, having a short wavelength absorption edge at about 200 nm and a long wavelength edge at over 50 microns. Diamond is extremely hard and insoluble and apart from the presence of C stretch bands in the mid-infra-red is a near perfect optical material.

Efforts have been made over a number of years to fabricate both diamond films and substrates. The earliest progress was made to deposit amorphous carbon ($\propto$ $-$C:H) or diamond like carbon (DLC). This was done successfully in a number of centres (Stein 1981, Bubenzer 1981) and was vigorously promoted as the best solution to both anti-reflectance coating and hardening of Ge windows for use in the infra-red. It has been shown that the refractive index can be tuned from about 1.9 to 2.3 by varying the amount of hydrogen used in the plasma deposition process. Unfortunately, although DLC coatings are hard they oxidise in the presence of oxygen at moderately high temperatures thus limiting their usefulness as thermal imager protective coatings. DLC coatings are grey in colour and only transparent in the far-infra-red. In addition, although the refractive index can be tuned so that an exact anti-reflectance coating ($n = 2$) can be deposited on a germanium substrate ($n = 4$) the infra-red absorption is high (i.e. a $\lambda/4$ DLC a-r coating at $\lambda = 10.6\,\mu\text{m}$ has a 3%A) thus vitiating its usefulness with high power CO_2 lasers.

More recently intense research has been carried out on the production of diamond ($n = 2.4$) coatings and substrates. There is a world wide interest as can be seen from the proceedings of the SPIE Conference in San Diego (e.g. Klein 1992, Harris 1992) and the 1993 Diamond Conference in Portugal (Sussmann 1993a). Diamond coatings are usually fabricated using the chemical vapour deposition, CVD, process whilst substrates can be produced using either CVD or an oxy-acetylene torch process (Sussmann 1993a, Brierley 1991). CVD grown diamond is polycrystalline and depending on the precise deposition conditions can have good transparency (Wort *et al.* 1993), high thermal conductivity (Sweeney *et al.* 1993) and high laser induced damage thresholds (Sussmann *et al.* 1993). The transparency is a function of the deposition conditions and the lower absorption cut-off edge can vary from ¯2 μm down to 200nm. Figure 4.3 shows (a) the u-v/visible/near i-r and (b) the infra-red transmission spectra of a typically clear polycrystalline diamond substrate (Sussmann 1993b).

4.5.2   *Nitrides*

Nitrides are perhaps the most neglected and also the most promising optical materials that are known to be available. The problem lies not in any way due to adverse properties but because of their general unreactivity and difficulty to grow. Although small crystals have been produced, and therefore the crystal class is known, the only widely available material is polycrystalline. In recent years the advent of micro-wave assisted plasma chemical vapour deposition has allowed the deposition of thin films of such nitrides as Si_3N_4, Ge_3N_4, AlN, BN. These materials are expected to revolutionise the thin film industry as the combination of gaseous or liquid precursor and the MPACVD technique allows the deposition of graded index layers and ultimately true rugate design filters (see Wood *et al.* 1990, Greenham *et al.* 1993). In general the nitrides have a wide transparency band reaching from the ultra-violet into the infra-red. Each material has its own characteristic absorption band in the infra-red, dependent on the molecular weight of the anion, with a further transparency region at longer wavelengths. Table 4.1 gives a list of the optical and physical properties of a selection of these materials.

4.5.3   *Phosphides*

The optical properties of the phosphides are somewhat the same as the nitrides except that it does not seem to prove possible to grow mixed material with continuously variable refractive indices.

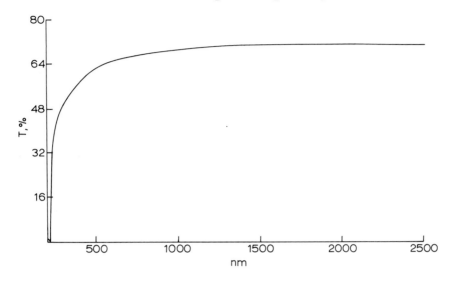

Figure 4.3   (a) Visible and (b) Infra-red transmission spectrum of polycrystal-line diamond

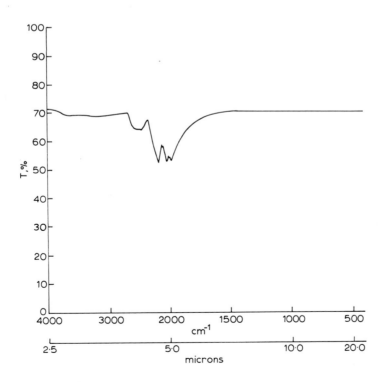

Table 4.1   Optical and physical properties of diamond,
diamond films and nitrides

Material	Spectral transmission range (μm)		Hardness kgm m^{-2}	RI (n)	MP (°C)
Diamond	0.2	→ >50	9000	2.42	>3500
Diamond film	1	→ 50	9000	2.4	>3500
Amorphous carbon film	8	→ 12	–	2.0	–
Ge–C	2	→ 11	700	2 → 4	–
B–P	0.6	→ 12.1	759	–	3200
GaP	0.45	→ 8.3	845	3.0	1467
Si$_3$N$_4$	0.27	→ 9.2	2200	2.05	2160
Ge$_3$N$_4$	0.3	→ 20	–	2.2	–
Al N	0.17	→ 19	2040	1.95	1700
Ti N	0.7	→ >2.5	–	1.7	2950
Zr$_3$N$_4$	0.4	→ 12.5	–	3.2	2980
Hf$_3$N$_4$	0.3	→ 12.5	–	2.1	3390
BN	0.2	→ 25	4500	2.1	3000

4.6   Polarising Materials

All anisotropic crystals are potentially birefringent, having different refractive indices in different crystallographic directions. As would be expected, there are families of crystals which exhibit high birefringence and it is possible to make an informed guess at the polarising potential of any given material especially if the crystal class and the lattice spacing are known. Table 4.2 lists a selection of the data for the best known birefringent materials. It can be commented that:

(a)   The *carbonates* have covalent bonding and yield many birefringent materials. Although calcium carbonate is the most common, and is widely used as a polarising material, there are several other carbonates with better parameters. Unfortunately these crystals tend to be difficult to grow, although lead carbonate has been grown using the hydro-thermal technique.

(b)   The *sulphides* grow as asymmetric crystals exhibiting appreciable birefrin-gence. However, these crystals are soft and suffer from absorption because of the relatively small band gap.

(c)   The *oxides*, having ionic bonding, are more likely to form cubic compounds but there are non-cubic crystal forms with useful indices. The majority of these crystals are colourless, non-absorbent and hard. Unfortunately they mostly form more than one structure and the preferred, or most easily grown, form is the one with the lowest birefringence.

Table 4.2   Data on the best known birefringent materials

Material	Chemical formula	Structure	n_e	n_o	Δ_n	Growth method
Calcite	$CaCO_3$	Hexagonal	1.48	1.66	0.17	Hydrothermal
–	$MnCO_3$	Rhombic	1.60	1.82	0.22	
–	$FeCO_3$	Trigonal	1.63	1.87	0.24	
–	$ZnCO_3$	Trigonal	1.62	1.82	0.20	
–	$SrCO_3$	Rhombic	1.52	1.67	0.15	
Cerussite	$PbCO_3$	Rhombic	1.80	2.07	0.27	Hydrothermal
Orpiment	As_2S_3	Monoclinic	2.40	3.00	0.66	
Proustite	$Ag_2SeAs_2S_3$	Hexagonal	2.79	3.08	0.29	
Pyrogyrite	$Ag_6Sb_2S_6$	Hexagonal	2.88	3.08	0.2	
Quartz	SiO_2	–	1.53	1.54	0.01	
Brookite	TiO_2	Rhombic	2.58	2.74	0.16	
Rutile	TiO_2	Tetragonal	2.62	2.90	0.28	Hydrothermal/Czechralski
Valentinite	Sb_2O_3	Rhombic	2.18	2.35	0.17	
Tellurite	TeO_2	Rhombic	2.00	2.35	0.35	Hydrothermal
Para-tellurite	TeO_2	Tetragonal	2.27	2.43	0.16	
Geikelite	$MgTiO_3$	Rhombic	1.95	2.31	0.36	Hydrothermal
–	$MnTiO_3$	Triagonal	2.20	2.47	0.27	Hydrothermal/Czechralski
–	YVO_4	Tetragonal	2.00	2.23	0.23	Hydrothermal/Czechralski

(d)   Crystalline *organic salts* often exhibit high birefringence particularly when they grow in flat platelets or needles. Unfortunately there are very few data as to the refractive indices of these materials.

Dichroic crystals form another class of materials used for polarisation. These crystals, which have an extremely well defined optic axis, selectively transmit the polarisation component parallel with this axis and absorb that at 90°. Dichroism is exhibited by a number of minerals and also by a large number of organic compounds. In recent years this has been achieved by packing thin sheets of an organic polymer with ultramicroscopic polarising crystals and stretching the combination so that the optic axes become parallel. For instance, Polaroid is manufactured by stretching polyvinyl alcohol films (to line up the long strings of molecules) and then impregnating the film with iodine (Land 1951).

CHAPTER FIVE

Optically Non-Linear Materials

5.1 Electro-Optic Effect

The structure of the unit cell (both size and atomic or ionic arrangement) determines the crystal structure and the resulting symmetry or asymmetry similarly determines the physical properties of the material. The size and arrangement on the atomic scale therefore has a direct influence in determining the susceptibility and non-linear properties of a given material. For example, centro-symmetric crystals – crystals which exhibit inversion symmetry – do not display second-order non-linear effects such as harmonic generation but as the fourth-rank susceptibility tensor does not reduce to zero third-order non-linear effects can occur. Table 5.1 lists the susceptibility terms along with their associated optical effects and main applications (Kolinsky and Jones 1989). Good reviews of the subject have been published by Yariv (1984), Butcher and Cotter (1990) and Agrawal and Boyd (1992).

5.1.1 First-Order Effects

The first-order term is responsible for refraction and is the physical basis for all low power physical optical effects.

Rayleigh scattering is a first-order effect. It is caused by thermally arrested density fluctuations in the lattice; these result in localised changes in the dielectric constant. The effect is $\tilde{}\omega^4$ and therefore principally affects the high frequency region of the spectrum (ultra-violet as opposed to the infra-red).

5.1.2 Second-Order Effects

The susceptibility, χ, is a tensor which can in turn be expressed as a matrix made up of the relevant non-linear susceptibility coefficients, d_{im}. Table 5.2 lists the second-order non-linear susceptibilities of some of the main non-linear crystals (Al-Saidi and Harrison 1985, Lin and Chen 1987, Kolinsky and Jones 1989, Eckart *et al.* 1990, Higgins 1992). Large values of d_{im} denote more efficient non-linear behaviour. It will be realised from inspection of this

Table 5.1   Applications of optical non-linear effects

Susceptibility term	Effect	Application
$\chi^{(1)}$ ($\omega = \omega$)	Refraction	Lenses Optical fibres Physical optics
$\chi^{(2)}$ ($2\omega = \omega + \omega$)	Frequency doubling	Second harmonic generation
$\chi^{(2)}$ ($\omega_s = \omega_a \pm \omega_b$)	Frequency mixing	Frequency up-converters Optical parameter oscillators (down-converters) Spectroscopy
$\chi^{(2)}$ ($\omega = \omega + o$)	Electro-optic effect	Pockels cell Q-switch Phase/amplitude modulators Beam deflectors
$\chi^{(3)}$ ($3\omega = \omega + \omega + \omega$)	Frequency tripling	Third harmonic generation Spectroscopy
$\chi^{(3)}$ ($\omega = \omega + o + o$)	DC Kerr effect	Variable phase retardation Material investigation
$\chi^{(3)}$ ($\omega_a = \omega_a + \omega_b - \omega_b$)	AC Kerr effect Optical Kerr effect Raman scattering Brillouin scattering	Fast switching Time resolved (gating) experiments Generation of different wavelengths
$\chi^{(3)}$ ($\omega = \omega + \omega - \omega$)	Intensity dependant RI Self-focussing Degenerate four-wave mixing	Optical bistability Optical limitation Phase conjugation Real-time holography

table that the susceptibility coefficients have a significant relationship with both the molecular and the crystal structures.

Harmonic generation. If the light wave is written as:

$$E = E_\omega \cos(\omega t)$$

then the quadratic term yields a component at twice the fundamental frequency (2ω) and the cubic term at three times the fundamental (3ω).

$$\text{as } E^2 = E_\omega^2 \cos^2(\omega t) = \tfrac{1}{2}E_\omega^2[1 + \cos(2\omega t)]$$

$$\text{and } E^3 = E_\omega^3 \cos^3(\omega t) = \tfrac{1}{4}E_\omega^3[3\cos\omega t + \cos(3\omega t)]$$

This radiation is termed second and third harmonic radiation respectively and is one of the major applications, being used to generate a variety of wavelengths where suitable fundamental sources are unavailable.

It is possible to pass a fundamental wavelength, λ_1, into a non-linear electro-optic crystal, operating under the right conditions, and to generate the

Table 5.2 Non-linear susceptibilities

Material		Transparency Range (nm)		Point Group	d_{im}	$pm\,V^{-1}$	Phase matching Range	(nm)	Phase matching Type
Quartz	SiO_2	250	1250	32	d_{11}	0.4	—	1250	—
					d_{14}	0.008	—	—	—
$Ba_2NaNb_5O_{15}$	—	—	—	2 mm	d_{31}	-14.7	—	—	—
					d_{32}	-14.7	—	—	—
					d_{33}	-20.1	—	—	—
$LiNbO_3$	—	400	5000	3 m	d_{22}	3.1	800	5000	—
					d_{31}	5.9	800	5000	—
					d_{33}	41	800	5000	—
$BaTiO_3$	—		—	4 mm	d_{15}	-17	—	—	—
					d_{31}	-18	—	—	—
					d_{33}	-6.7	—	—	—
$NH_4H_2PO_4$	(ADP)	200	1500	$\bar{4}2\,m$	d_{14}	0.5	—	1500	—
					d_{35}	0.5	—	—	—
KH_2PO_4	(KDP)	200	1500	$\bar{4}2\,m$	d_{14}	0.5	517	1500	I
					d_{35}	0.46	732	1500	II
KD_2PO_4	(KD*P)	200	1500	$\bar{4}2\,m$	d_{14}	0.53	300	1300	—
					d_{36}	0.53	—	1300	—
$LiIO_3$	—	300	5500	6	d_{35}	-5.4	570	5500	I
					d_{32}	-4.2	—	—	—
LiB_3O_5	(LBO)	160	3500	2 mm	d_{15}	0.7	555	2500	—
					d_{32}	1.16	—	—	—
$\beta\text{-}BaB_2O_4$	(BBO)	189	3500	3 m	d_{11}	1.9	400	3300	I
					d_{22}	1.6	526	3300	II

Table 5.2   (*continued*)

Material	Transparency Range	(nm)	Point Group mm²	d_{im}	$pm\,V^{-1}$	Phase matching Range	(nm)	Phase matching Type
KTiOPO₄ (KTP)	350	4000		d_{15}	6.1	1000	2500	II
				d_{24}	7.6	–	–	–
				d_{31}	6.5	–	–	–
				d_{32}	5.0	–	–	–
				d_{33}	13.7	–	–	–
LAP	220	1950	–	–	20	440	1950	I
Urea	210	1400	–	d_{36}	8	473	1400	I
						600	1400	II
DAN	430	2000	–	–	2500	860	2000	–
				d_{eff}	27	–	–	–
NPP	500	2000	–	d_{21}	102	–	2000	–
PNP	500	2000	–	d_{21}	48	–	2000	–
POM	414	2000	–	d_{14}	170	830	2000	I
MAP	472	2000	–	d_{21}	800	900	2000	I
MNA	400	2000	m	d_{11}	250	–	2000	–
				d_{12}	38	–	–	–
mNA	500	2000	2 mm	d_{14}	30	1000	2000	I
CdS	–	–	6 mm	d_{31}	37.7	–	–	–
				d_{33}	36	–	–	–
				d_{36}	42	–	–	–
CdSe	750	20000	6 mm	d_{33}	50	–	20000	–

Table 5.2 (*continued*)

Material	Transparency Range	(nm)	Point Group	d_{im}	pmV^{-1}	Phase matching Range	(nm)	Phase matching Type
Proustite · Ag₃AsS₃	630	13000	3 m	d_{22}	28	–	13000	–
				d_{31}	15	–	–	–
CdGeAs₂	2400	18000	$\bar{4}2\,m$	d_{36}	457	–	18000	–
AgGaS₂	500	13000	$\bar{4}2\,m$	–	14	–	13000	–
AgGaSe₂	710	18000	$\bar{4}2\,m$	d_{36}	34	–	18000	–
Pyrargyrite AgSbS₃	–	–	3 m	d_{31}	12.6	–	–	–
				d_{32}	13.4	–	–	–
ZnGeP₂	740	12000	–	–	–	–	12000	–
Te	3800	32000	–	d_{11}	5.4	–	32000	–
HgIn₂Te₄	1400	40000	$\bar{4}2\,m$	d_{36}	1.4	–	40000	–

second harmonic, λ_2, such that $\lambda_1 = 2\lambda_2$. This occurs when the assemblage of induced dipoles oscillates coherently (i.e. with a definite phase relationship) so that their radiation fields add together constructively. The condition under which this occurs is termed phase matching and the general condition which has to be satisfied is:

$$k_1 = 2k_2 + \Delta k$$

where $k_1 = n_1\omega_1/c$, $k_2 = n_2\omega_2/c$

where n_1 is the refractive index at frequency ω_1, wavelength λ_1
$\qquad n_2$ is the refractive index at frequency ω_2, wavelength λ_2

The crystal momentum wavelength, k, is related to the refractive index and consequently to the direction of propagation through the crystal with respect to the optic axis. The second harmonic power generated by a single Gaussian beam of frequency, ω_1, and of power, P_ω, incident on a plane parallel slab of thickness, L, of a non-linear crystal is given by:

$$P_{2\omega} = \frac{JP_\omega^2 L^2 d^2 \sin^2\theta [\sin(\Delta kL/2)]^2}{W_0^2 \qquad [\Delta kL/2]}$$

where J = constant
$\qquad d$ = SHG coefficient
$\qquad W_0$ = minimum beam radius
$\qquad \theta$ = angle between the crystal optic axis and the incident beam
$\qquad \Delta kL$ = phase mismatch between the fundamental and second harmonic.

In general, because of the crystal dispersion $\Delta k \neq 0$ and for a fixed Δk the function undergoes periodic oscillation as a function of the crystal length with a period of $2\pi/\Delta k$.

When $\theta = 90°$ the conversion efficiency is a maximum. This happens when the fundamental and second harmonic beams propagate normal to the optic axis. There are two specific angles for most crystals according as to whether the electric fields of the two waves have the same or different planes of polarisation (termed Types I and II). It is also possible to phase match by temperature tuning as the refractive indices usually vary with temperature.

The intensity required to induce the non-linear process is reduced further, by many orders of magnitude, if the optical frequency lies close to a resonant frequency of the oscillating dipoles (this is termed resonant enhancement).

Under the right conditions it is also possible to generate third and higher harmonics of the fundamental wavelength.

Pockels effect. In the case where there is both an oscillating electromagnetic field and an applied d.c. electric field the plane of polarisation will be rotated (Pockels effect). If the applied electromagnetic wave is plane polarised then the combination can be used as a polarisation rotation mechanism or switch. This effect is used to great effect in the design of high power laser systems as the electric field can be applied very quickly (Pockels cell). Useful reviews of the effect are given by Yariv (1984) and Kolinsky and Jones (1989).

The total retardation produced by a crystal can be expressed as

$$T = T_0 + dT(V)$$

where T_0 is the retardation caused by the natural birefringence of the crystal.

$dT(V)$ is the retardation induced by the applied electric field V. The natural birefringence is usually cancelled out by arranging the optical path to be a double path, reflecting back on itself and the transmitted light is therefore controlled by the electrically induced birefringence.

As the symmetry of the crystal determines which of the electro-optic coefficients (r_{13}, r_{33} etc.) are non-zero there is a specific relationship between the crystal structure and the operation of the Pockels cell. For example, for a crystal with trigonal 3 m symmetry (such as lithium niobate), with the electric field applied in the z-direction and the light polarised at 45° to both the z- and y-axes and the incoming light propagating along the x-axis, the phase retardation will be

$$T = T_0 + dT(V)$$
$$= (\pi L/\lambda)\{2[n_e - n_0] + [n_e^3 r_{33} - n_0^3 r_{13}](V/d)\}$$

where d is the crystal width across which the voltage V is dropped and
$\quad\quad$ L is the crystal length.

A useful measure of the performance of such a cell is the voltage required to induce a phase retardation of π radians and hence to switch from full extinction to full transmission.

$$V_\pi = \frac{\lambda}{(n_e^3 r_{33} - n_0^3 r_{13})} \cdot \frac{d}{L}$$

Comparison is more readily achieved for the condition where $d/L = 1$, the 'reduced' half-wave voltage. A brief comparison of the half-wave voltages available is shown in Table 5.3.

Another class of electro-optic materials, the dihydrogen phosphates (commonly termed XDPs) crystallise in the 42 m point group configuration. These include ammonium dihydrogen phosphate (ADP), potassium dihydrogen phosphate (KDP) and potassium dideuterium phosphate (KD*P). These materials being uniaxial are more conveniently used in the longitudinal

Table 5.3   Electro-optic data for a range of materials

Material	Crystal Symmetry	$n^3 r (10^{-12} m/V)$		Longitudinal (L) or Transverse (T)	$V\pi$ kV at 633 nm
LiNbO$_3$	3 m	$(n_e^3 r_{33} - n_0^3 r_{13})/2$ Optimised cut	$= 112$	T L	2.94 2.50
SBN : 60	4 mm	$(n_e^3 r_{33} - n_0^3 r_{13})/2$	$= 1300$	T	0.25
SBN : 75	4 mm	$n_0^3 (r_{33} - r_{13} - 2r_{42})/2 \cos\theta$		L	0.10
BBO	R3c $\equiv$ 3m	$n_0^3 r_{yyy}$	$= 11.6$	T	56
KD*P	42 m	$n_0^3 r_{63}$	$= 90$	L	3.50
BaTiO$_3$	4 mm	$n_0^3 (r_{33} - r_{13} - 2r_{42})/2\cos\theta$		L	0.037
KTaNbO$_3$	4 mm	$n_0^3 (r_{33} - r_{13} - 2r_{42})/2\cos\theta$		L	0.005
mNA	2 mm	$(n_z^3 r_{33} - n_y^3 r_{23})/2$	$= 63$	T	5.0
MNA	Cc $\equiv$ m	$(n_1^3 r_{11} - r_3^3 r_{31})/2$	$= 270$	T	1.3
DCNP	Cc $\equiv$ m	$n_2^3 r_{333}/2$	$= 860$	T	0.37
SPCD		$n_z^3 r_{33}/2$	$= 800$	T	0.5

configuration, where the field is applied in the same direction as the light propagation. In this case the induced phase shift is:

$$T = 2\pi n_0^3 r_{63} V/\lambda$$

and is therefore independent of the crystal path length, L. The half-wave voltage is then given by the formula

$$V_\pi = \lambda / 2n_0^3 r_{63}$$

Inspection of the values reproduced in Table 5.3 indicates that some of the inorganic crystals should have even lower half-wave voltages than the best of the long-chain organic molecules (Yazaki *et al.* 1971, Meredith 1986, Gorton 1986, Nayar and Winter 1990). Unfortunately these values have only been measured on very thin crystals and the geometry involves operation at an angle. This means that until the crystal growing problem is solved, and this is relevant for both the mixed niobates and the organic crystals, lithium niobate will be the most used optical switch except for the use of KD*P etc. where the use of sealed units is not a disadvantage.

5.1.3   Third-Order Effects

Brillouin scattering. Brillouin scattering occurs when the incident photon is scattered by an acoustic phonon. This effect occurs at low optical power due to thermal excitation but becomes increasingly serious at high optical powers as stimulated Brillouin scattering (SBS) transfers energy to the backward-travelling wave. SBS is especially serious in single mode fibre (see Chapter 8) (where is occurs at input power levels of a few mW) and measurements of the reflected fraction have been as high as 60%. The SBS threshold, P_B may be written as:

$$P_B = 21 . A/g_B L$$

where A = beam cross-sectional area
g_B = Brillouin gain coefficient
L = sample length
~ = 1.2 mW for a 10 μm diameter fibre (Stolen 1979)

SBS can be detrimental in a number of ways: by introducing additional attenuation, by causing multiple frequency shifts and by increasing backward coupling into the light source. In telecommunication applications it is particularly limiting in that it attenuates the power available for coherent detection techniques. In high power transmission applications the main problem is straightforward attenuation leading to enhancement of the power loading and lowering of the laser induced damage threshold.

Raman scattering. Raman scattering involves either the incident light photon emitting a phonon and re-radiating the remaining energy as another photon or a phonon being absorbed as the incident photon is converted to the scattered photon. Again the stimulated process results in a serious limitation of the power handling capacity of the optical fibre in telecommunication applications where the Raman threshold effectively marks the limit at which incoherent detection techniques can be used. The shift to lower frequencies of the forward Raman travelling wave can also result in added dispersion in the fibre. The primary effect may or may not be deleterious in the high power handling scenario where the total energy is only slightly modified. A list of Raman shifted wavelengths in fused silica is given in Table 5.4 (Pini *et al.* 1983).

The stimulated Raman threshold, P_R, may be written as:

$$P_R = 16.A/g_R L$$
$$\tilde{\ }440 \text{ mW for a 10 μm diameter fibre (Stolen 1979)}$$

where g_R = Raman gain coefficient
L = effective interaction length = $(1 - \exp(-\alpha l))$
where α is the attenuation coefficient and l is the sample length

5.2 Acoustic-optic Materials

When a frequency is imposed on a piezo-electric transducer bonded onto the

Table 5.4   Raman shifted wavelengths in fused silica

Order	Wavelength (A°)	Shift (cm^{-1})
Fundamental	3080	–
S1	3124	463
S2	3167	431
S3	3208	403
S4	3253	430
S5	3299	428
S6	3347	434
S7	3395	428
S8	3445	428
S9	3491	383

Pini *et al.* (1983)

side of a block of material (see Figure 5.1) an acoustic vibration is induced in the material and a standing wave is induced. These acoustic waves result in refractive index changes, with the periodicity of the acoustic wave, such that a beam of light passing through the glass is diffracted. The induced polarisation or perturbation varies spatially with a periodic length equal to the acoustic wavelength. The effective frequency/wavelength of a beam of light passing through the crystal can be both increased or decreased.

$$v = c/\lambda \pm mf$$

where λ is the wavelength of the probe beam
f is the transducer frequency
m is an integer

Figure 5.1 illustrates the phenomena. Conservation of momentum causes the different shifted frequencies to propagate at different angles (the Bragg condition, θ), i.e. $\theta = mf(l/v)$ where v is the velocity of the acoustic wave through the crystal.

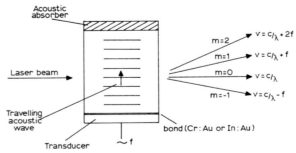

Figure 5.1   Acousto-optic cell schematic

The intensity of the different orders with respect to the unshifted zero order is a function of the acoustic wave power density and the electro-optical coefficients of the crystal. These elasto-optical coefficients determine the efficiency with which an acoustic wave can perturb the optical properties of a crystal, in particular the inverse dielectric tensor. Under certain conditions higher orders can be suppressed so that most of the intensity is diffracted into the first-order beams ($m = 1$).

Normal technology limits the choice of acousto-optic material to LiNbO$_3$, TeO$_2$, quartz or to extra dense flint glasses in order to achieve relatively high diffraction efficiencies. The cells themselves are fairly simple, usually comprising a cube of acousto-optic material with a transducer bonded onto one side. (lead zircotantalate, lithium niobate, strontium barium niobate (SBN), ferroelectrics) and an acoustic damping layer bonded to the other to reduce reflections and complicated ripple interference patterns. Triangular cells with transducers on each of the side faces have been constructed to achieve ultra-fast switching with minimal reflections for laser Q-switch operation.

Acousto-optic devices can be used in three configurations:

(1) Amplitude modulation is produced by using the first-order diffracted output beam and varying the RF (acoustic wave) power. This can also be used as a switch.
(2) Beam deflection is effected by using the diffracted output beam and varying the RF frequency.
(3) A laser beam with frequency side-bands can be produced by recombining the undiffracted and the first-order beams.

The switching speed of the effect depends on the transit time of the acoustic wave across the optical beam and can therefore be increased by focusing the laser beam down to a small spot size. Commonly available devices have switching speeds ranging from 10^{-5} to 10^{-8}s.

The ability to rapidly modulate the amplitude and direction of a laser beam by simply applying an RF field to a block of suitable material has resulted in the widespread use of acousto-optic technology throughout the fields of reprographics, imaging, laser Q-switching and the laser entertainment industry. A true two-dimensional deflection can be gained by simply using two thin, large aperture acousto-optical cells orthoganal to each other.

By generating two optical beams (input and first-order) and redirecting them it is possible to produce an interference pattern. This fringe pattern is not stationary, as there is an RF frequency shift, so the pattern moves in the direction perpendicular to the fringes at the RF frequency. This phenomenon has been used to measure the particle flow in a medium sited at the point of beam recombination. Stationary particles generate a pulsed scatter signal at the RF frequency and a moving particle generates a signal that is Doppler

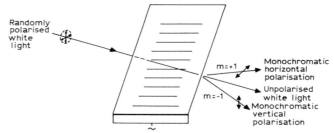

Figure 5.2   Acousto-optic tunable filter

shifted to higher or lower frequencies depending on its direction of motion relative to the fringe orientation. This allows both the speed and direction of the flow to be monitored. When crystalline acousto-optic material is used the orientation with respect to the transducer is important whereas this is not necessary for glass. The advantage some crystalline materials have over glass is that the best materials have higher refractive indices and have higher Bragg angles.

Acousto-optic tunable filters (AOTF) can be fabricated if the acousto-optic material is also birefringent. Figure 5.2 shows the schematic of a non-colinear cell (the acoustic and optical beams being at different angles to each other). As in the ordinary Bragg A–O modulator, momentum conservation dictates that the frequency shifted beam undergoes a direction change, as illustrated. In an AOTF the collimated incident light interacts with the acoustic wave over a relatively long distance and the frequency shift is analogous to harmonic generation and frequency mixing in optically non-linear materials. Because of the coherence length of the interaction a phase matching condition is required to avoid the net intensity of the shifted light summing to zero because of destructive interference. This is done by phase matching using a birefringent material (such as TeO_2). Since the shifted light undergoes a $\pi/2$ polarisation change there is therefore only one wavelength at which the shifted and unshifted light have the same phase velocity and therefore only one wavelength will be diffracted into the first-order beam. As the phase-matched wavelength varies with the acoustic frequency the AOTF can therefore be used as a tunable narrow-band filter.

If an AOTF is used with unpolarised white light it effectively resolves the incoming light into parallel ($m = +1$) and perpendicular ($m = -1$) polarisations, the phase matching condition causing one polarisation to be upshifted and the other downshifted. The dispersion and bandwidth depends on the degree of collimation of the light beam and the length of the acousto-optic cell.

CHAPTER SIX

Laser Materials

6.1   Inversion, Fluorescence and Stimulated Emission

The origin of the optical absorption bands was discussed in Chapter 2. At zero degrees K all the electrons in the electron cloud surrounding the nucleus are in their ground states – the lowest energy levels. At temperatures above absolute zero the electrons in the upper levels – the conduction bands – are continually excited and deactivated between the adjacent empty levels. When a photon of the correct energy impinges upon the material the energy is transferred to an electron in the ground state. This electron is now energised into an orbit with a higher potential energy, $E = h\nu$ (where h is Plank's constant). The electron will stay in this excited state for a period of time but will eventually relax back down to the ground, or some intervening, state. The lifetime of the electron in the excited state is termed the fluorescent lifetime, t, and can vary between 10^{-14} and 10^{-2}s. The electron can lose energy by radiative (emission of light) or by non-radiative (emission of phonons) means. The former process is termed fluorescence, phosphorescence or laser action, while the latter is thermal. The former leaves the material undisturbed whilst the latter leaves it hotter. The relaxation paths may be a mixture of radiative and non-radiative processes and can also be spontaneous or stimulated. Spontaneous relaxation is termed fluorescence and has an associated lifetime, t. If this lifetime is long enough then it is possible, once there are more electrons in the excited state than in the terminal state, for a photon of the appropriate wavelength to stimulate an electron out of the upper level and down to the lower level with an accompanying emission of photonic energy. The terminal state may or may not be the ground state.

   The excitation may be optical (phosphors and lasers) or electrical (electroluminescence, light emitting diodes and semi-conductor lasers). In general terms the former materials absorb short wavelength, high energy photons and emit longer wavelength, lower energy photons and phonons whilst the latter convert electrical charge into photonic energy. The pump energy may be either coherent or incoherent, monochromatic or multi-wavelength. Fluorescence and electroluminescence are incoherent and are of a broad band emission. Lasing action is coherent and monochromatic.

6.2   Optically Driven Emission

Laser (Light Amplification by Stimulated Emission of Radiation) Action is a physical optics phenomena which occurs in a host of materials but was not predicted until 1958 by Schawlow and Townes and not demonstrated until 1960 by Maiman.

Lasing action can occur in any suitable material which hosts atoms with suitable energy level schemes, the sole criterion being whether the electrons stay in the upper level long enough to allow inversion (more electrons in the upper lasing level than in the terminal level) to occur. The material can be glassy, crystalline, semi-conducting, liquid or gaseous. The following sections will discuss lasing action in a selection of the most common solid and solid-state materials but it should be remembered that the physics is the same whether the material is solid, liquid or gaseous (Goodwin and Heavens 1968).

The laser process may be understood relatively easily if we start with the generalised 4-level model of Nd^{3+} ions (see Figure 6.1). The pump light raises electrons from the ground state (1) to the excited state levels (4). These

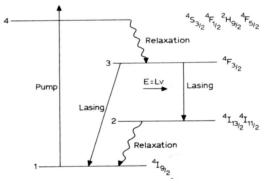

Figure 6.1   Generalised 4-level model for Nd^{3+} ions

electrons then decay to intervening levels (3) and (2). The transition (4) → (3) is a non-radiative transition (the energy is imparted in the form of a lattice vibration-phonon). These transitions are very fast. If the fluorescence lifetime of the electron in level (3) is relatively long then a population inversion will build up and conditions may be right for stimulated emission to take place. This occurs when a photon of the correct frequency ($E = h.v$) passes through the crystal and stimulates another photon causing the electron to lose energy and to drop from level (3) to level (2). The electron will then drop immediately from level (2) to level (1) by emission of a phonon. The stimulated electron will naturally be in phase with the stimulating electron and

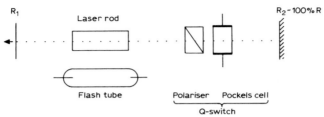

Figure 6.2   Solid state laser resonator schematic

so the laser beam is said to be coherent. In order for a significant amount of stimulation to occur a high transmission path must be present. This is obtained by fabricating the laser material in rod form and arranging the photons to be reflected back along the same path by providing high reflectance mirrors arranged parallel to each other (see Figure 6.2). In order for a fluorescent material to exhibit laser operation the round trip optical gain must exceed the round trip loss in the cavity. If we consider a laser rod of length l, mirror end reflectivities R_1 and R_2, round a laser host medium with a gain coefficient of β in a resonator with loss $(1-L)$ and scatter $(1-S)$ then the condition for light amplification by stimulated emission occurs when the round trip gain of photons in the cavity, $\exp(2l\beta N)$, exceeds the round trip loss $1/RLS$ (where $R = R_1 R_2$)

$$\text{i.e. } \exp(2l\beta N) > 1/RLS$$

In this equation N is the inversion, $(N = N_3 - N_2)$, the number of electrons in level (3) (the upper lasing level) minus the number in level (2) (the lower lasing level). This number has to be positive before stimulated emission can occur.

The gain coefficient, β, is a function of this inversion, N, the linewidth of the transition, Δv, the spontaneous transition probability, A_{32}, (the probability of an electron dropping from level (3) to level (2) spontaneously due to fluorescence) and the refractive index n.

$$\beta = \frac{\lambda^2 A_{32} N}{8\pi n^2 \Delta v}$$

In a crystalline material the energy levels are discrete and the line width of the transition, Δv, is narrow. In an amorphous material the effective energy levels are broader (even though the atomic widths are the same) and the line width is relatively broad. The effect of this can be seen by comparing the Nd^{3+} system in YAG (Kushida 1968) and silicate glass (Young 1969). The former material has a narrow line width, a low energy threshold for lasing, a high gain and a low saturation output energy. Nd : glass on the other hand shows relatively wide line widths, appreciably higher lasing thresholds and have far greater energy storage capability. These characteristics dictate the different usages of the two host systems. For instance, the large laser fusion systems start with Nd : YAG oscillators (for short pulse width, frequency control and low threshold) and then graduate to Nd : fluorophosphate glass amplifiers to generate the extremely high energies per pulse required for the application.

The flash tube input energy, E, is associated with the gain condition and the pumping efficiency, f, by the formula

$$E = \frac{N_3 h\nu\, V}{f}$$

where V is the volume of the laser rod

and E is the input energy required to pump N_3 electrons from the ground state to the upper lasing level in order to obtain an electronic inversion, N, despite a proportion of the electrons decaying to the lower lasing level spontaneously, i.e.

E_T (the input energy for the laser threshold)

$$= \frac{N_3 h\nu\, V}{f} = \frac{h\nu r^2 \ln RLS}{f 2\beta}$$

and the energy output, E_{out}

$$= F(E_{IN} = E_T)$$

where the slope efficiency, $F = f\eta$
where f is the electronic pumping efficiency
and η is a function involving the resonator losses (specifically R,L,S).

The electrical to photonic pumping efficiency, f, is given by the product of the factors:

	Approximate values
Conversion efficiency (electrical to light) of the flash lamp	0.25
Transfer efficiency of the pumping chamber	0.7
Fraction of the lamp spectrum absorbed by the lasing atom	0.1
Quantum efficiency $h\nu_{laser}/h\nu_{pump}$	0.7
Fluorescent efficiency of the upper lasing level	0.7
Number of excited electrons available at the time of Q-switching	0.7
	0.006

This gives a value of f of around 5×10^{-17} Joules/photons.

However, this value must also be modified as the pumping usually leads to a non-uniform radial distribution of the pump light, stress-induced birefringence and the crystal to areas (or filaments) of higher and lower transmission.

For this reason the effective pumping efficiency, f, may be either higher or lower than the theoretical efficiency calculated above.

$1-L$ is the energy loss in the resonator due to causes other than the mirror losses and the crystal scatter loss. It is affected by diffraction losses and by reflectivity losses from other optical components in the resonator. The diffraction loss is related to the size of the lasing filament and this changes within the lasing cycle as well as being a function of the flash-tube input. The reflectivity losses from the other optical components can be fairly large unless great care is taken to coat them all with efficient anti-reflecting coatings. This is because, especially when the laser is to be Q-switched, a phenomenon termed super-radiance occurs at high input energies. This is fundamentally the same as normal lasing (usually termed 'free lasing') except that the oscillations build up between surfaces other than the mirror surfaces. Typical values for this factor are of the order of 0.7.

6.3 Specific Laser Materials

6.3.1 Ruby

Ruby is a crystalline form of aluminium oxide doped with chromium, Al_2O_3 : Cr^{3+}. Ruby crystal occurs naturally and is valued highly as a precious stone not least because of its high refractive index (and therefore sparkle) and its hardness. It can be grown artificially by both the Verneuil and the Czochralski methods in large boules suitable for 6-inch long $\times$ 1-inch diameter laser rods to be fabricated. Al_2O_3 crystallises in hexagonal form and is therefore birefringent (see section 3.4). This is made use of in practical laser systems as it ensures that the laser beam is polarised (this is necessary if short pulses – Q-switching – are required). In practice the laser rod is fabricated so that the optic axis is at right angles to the axis of the rod (the direction the laser beam is generated) and the laser rod is aligned in the laser cavity (pumping chamber) so that the maximum absorption is gained, the optic axis and the flash-tube being in the same plane.

The ruby laser is a three-level system (see Figure 6.3), the terminal level being the same as the ground state. Because of this the laser is inefficient unless either the ground state is partly depopulated (by cooling the crystal below 10°C) or by pumping the laser hard so that nearly all the electrons are in the upper lasing level. The electronic inversion is

$$N = \frac{N_2 - N_1}{N_0}$$

where $N_0 = N_1 + N_2$ and is the total number of Cr^{3+} ions in the laser rod. As stimulated emission cannot occur until there are more electrons in the upper

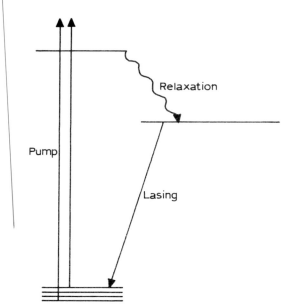

Figure 6.3   Generalised 3-level model for ruby ($Al_2O_3:Cr^{3+}$)

lasing level (2) than in the lower lasing level (1) (i.e. $N > 0$ the threshold for lasing occurs when $N_2 = N_1$ and the maximum output occurs at twice the threshold energy when $N_2 = N_0$ (see Figure 6.4).

It is therefore possible to derive that the threshold energy is given by

$$E_T = \frac{N_0.hv}{2f}$$

The maximum output by $E_{SAT} = N_0.hv.F$ and the input energy at which this output occurs by $E_S = 2E_T = N_0.hv/f$ where $N_0 = VC$

where V is the volume of the laser rod
      C is the Cr^{3+} concentration, atoms.cm^{-3}
      f is the pumping efficiency
      F is a function of the resonator

It will be seen from these calculations that the lasing threshold is a function of the Cr^{3+} doping level and the lamp efficiency but not the mirror reflectivities or the resonator losses. The input energy at which the maximum output is gained is twice the threshold energy. The maximum output energy which can be extracted from a ruby laser rod is a direct function of the Cr^{3+} concentration and the cavity losses but is independent of the pumping efficiency (Randle and Stitch 1969).

For most systems F (the resonator efficiency) has a value $0.5 < F < 1$,

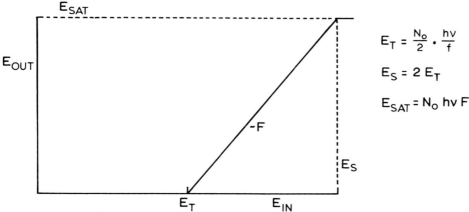

Figure 6.4 Ruby laser energy input/output characteristic

$$E_T = \frac{N_o}{2} \cdot \frac{hv}{f}$$

$$E_S = 2\, E_T$$

$$E_{SAT} = N_o\, hv\, F$$

depending on the mirror reflectivities, the crystal quality and the Q-switching losses. The usual doping level of Cr^{3+} in ruby is 0.05 atom %, in practice varying between 0.03% and 0.01%. It is therefore possible to calculate the optimum size of laser rod for a given energy output together with the necessary energy input. For example, in order to allow an output of 100 mJ to be extracted from a ruby laser rod the alternatives are listed in Table 6.1. Other considerations which have to be taken into account are: that flash-tubes are more efficient the longer the length of the arc; that the beam divergence of the laser output improves with the length of the laser rod; that the pulse length increases with the number of transits of the laser resonator – the pulse length decreases as the total reflectivity of the mirrors decreases. One last imponderable is that it is difficult to make the whole of the volume to contribute to the lasing process at any time due to stress birefringence and lensing when the laser crystal is pumped hard. For these reasons it is normal to use output reflectivities of ˜70%R and to use the longest and thinnest rods

Table 6.1   Relationship between Cr^{3+} doping level and resonator efficiencies: ruby

Resonator Efficiencies	Cr^{3+} doping level, C% atoms cm^{-3}			E_T joules	E_{IP} per 100 mJ$_0$/P joules
	0.03	0.05	0.10		
0.5	0.144	0.086	0.044	30	60
0.8	0.090	0.054	0.028	19	38
1.0	0.072	0.043	0.022	15	30

The table shows the volume of the $Al_2O_3 : Cr^{3+}$ laser rod (cm^3) necessary to enable an output of 100 mJ to be extracted.
(nb if $r = 1$ mm then volume = $0.03 /cm^3 /cm$ length)

Table 6.2   Wavelength of operation: Nd^{3+} in a variety of host lattices

Host lattice		Wavelength ($\lambda\,\mu$m)
Phosphate glass	NdP_5O_{11}	1.050
Fluorophosphate glass		1.054
Calcium tungstate	$CaWO_4$	1.058
Silicate glass	SiO_2	1.060
Yttrium aluminium garnet	$Y_3Al_5O_{12}$ (YAG)	1.064
		1.3167
		1.3203
		1.3335
		1.3351
		1.3381
		1.3418
		1.3533
		1.3572
		1.4150
		1.4271
		1.4320
		1.4440
Yttrium aluminium oxide	Al_2O_3 (YALO)	0.930
		1.0645
		1.0729
		1.0795
		1.0909
		1.0989
		1.3371
		1.3411
Lithium niobate	$LiNbO_3$	1.0845

consistent with strength. In practice a fairly short, thin rod (say 20 mm by 1 mm radius) would be required to ensure an output of 100 mJ with an electrical input of 40 to 60 Joules, depending on the resonator.

6.3.2 Nd^{3+} : YAG

The most widely used laser ion is neodymium, Nd^{3+}, which has been made to lase in a whole variety of host lattices (see Table 6.2). The precise doping concentrations, absorption spectra, gain coefficients and lasing wavelengths vary from host to host. The specific 4-level model shown in Figure 6.5 is that pertinent to Nd^{3+} : YAG, the number of possible lasing lines indicating the possibility of high gain.

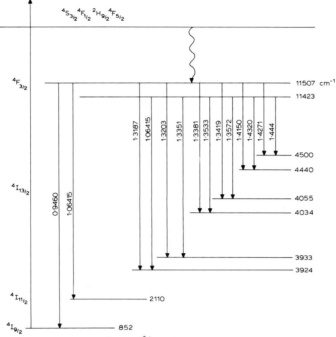

Figure 6.5 Laser transitions for Nd^{3+}: YAG

Figure 6.6 shows the transmission/absorption spectrum for a thin slice of Nd^{3+}: YAG crystal. This figure indicates that the main absorption bands (pump bands) occur in the red and near infra-red (600 and 850 nm). Unfortunately only a relatively small proportion of the pump light is absorbed

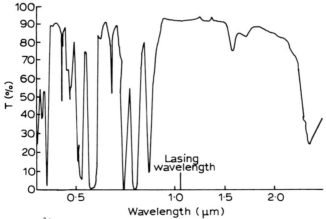

Figure 6.6 Nd^{3+} absorption spectrum

by the crystal (˜10%). As one electron is excited by each input photon the quantum efficiency can be calculated:

$$\text{Quantum efficiency} = h\nu \text{ laser}/h\nu \text{ pump}$$
$$= [(600 + 850)/2]/1064$$
$$= 0.7$$

Some of the electrons decay back to the ground state via other energy levels leaving a fluorescence efficiency of 0.7. Similarly it is unreasonable to expect all the inverted electrons to wait in the upper lasing level (the fluorescence lifetime is ˜240 µs) and this gives another loss and the percentage of excited electrons which are still available for Q-switching when the crystal is pumped with a flash pulse of ˜100 µs is 70%. These considerations lead to the calculation of the upper value for theoretical pumping efficiency as 1%.

It is possible to increase the pumping efficiency by one or other of the following means.

Increase the concentration of the Nd^{3+} ion. This poses a problem in the YAG host where concentrations of >1.5 atom% lead to unacceptable crystalline strain. Using the much more lenient glass hosts (see Table 6.2) it is possible to raise the Nd^{3+} concentration to over 5 atom% without distorting the lattice. This allows a greater percentage of the pump light to be absorbed.

Use additives to increase the fraction of the pump energy absorbed. There is a range of useful atoms, some of which lase themselves and some which allow their excited electrons to decay back down to the ground state via the energy levels of the lasing ions. Most of these are rare earths, such as Yb^{3+}, Ho^{3+}, Er^{3+} and Tm^{3+} which have approximately identical energy levels to the Nd^{3+} ions.

Neodymium doped yttrium aluminium garnet is grown from a heated crucible using the Czochralski technique. In this technique a mixture of yttrium oxide, aluminium oxide and neodymium oxide are mixed together and melted in a platinum crucible at a temperature of about 1900°C. An oriented crystal is then dipped into the melt, rotated slowly and gradually withdrawn. A photograph taken of a typical apparatus is shown in Figure 6.7. This figure also includes a photograph of a neodymium doped YAG boule and a jig being set up in order to polish several rods at the same time.

6.3.3   Other Laser Host Lattices

Nd^{3+}, Cr^{3+}, Ti^{3+}, Ho^{3+} and Er^{3+} are among the many ions that, in conjunction with suitable lattices, such as calcium tungstate, aluminium oxide, alexandrite, lithium niobate and yttrium fluoride have been proved to support lasing action. Some of the most important of these laser materials are listed in Tables 6.2 and 6.3. It will be seen that there are a whole range of wavelengths that can be generated and that the wavelength is decided by a

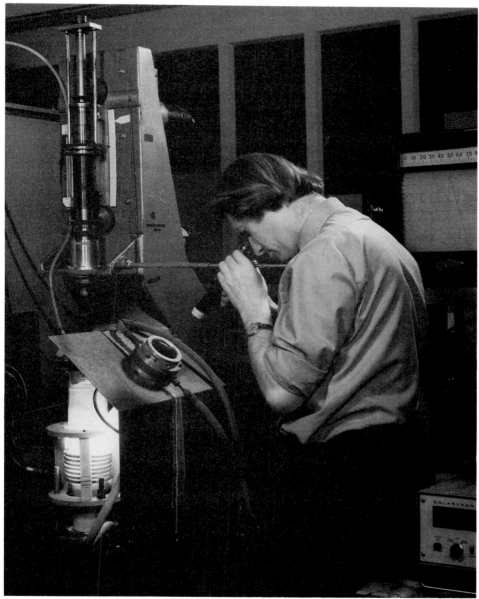

Figure 6.7   Neodymium-doped yttrium Oxide: Nd:YAG   (a)
                (a) Czochralski crystal growing apparatus
                (by kind permission of GEC-Marconi Ltd, Hirst Research Centre)

combination of the ion and the host lattice. The pumping efficiency of many
of these combinations is not always as high as Nd:YAG because the twin
combination of spectral absorption and quantum efficiency is often fairly low.

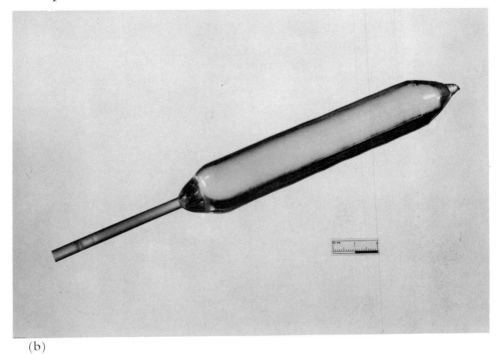

(b)

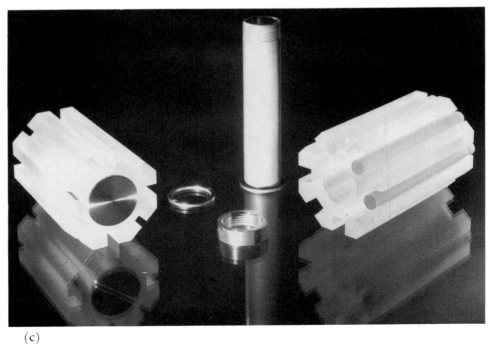

(c)

Figure 6.7   Neodymium-doped yttrium Oxide: Nd:YAG
(b) Nd : YAG crystal boule
(c) Polishing jig
(by kind permission of GEC-Marconi Ltd, Hirst Research Centre)

Table 6.3 Wavelength of operation of a variety of laser ions/host lattices

Host	ion	Wavelength of operation (nm)	Comments
Al_2O_3	Ti^{3+}	$675 \rightarrow 945$	Tunable
$BeAl_2O_4$	Cr^{3+}	$700 \rightarrow 800$	Tunable
Glass	Er^{3+}	1540	Eye safe
$LiYF_4$	Ho^{3+}	2060	Eye safe
YAG	Er^{3+}	2940	Eye safe

However, many of these materials can be made more efficient by the addition of sensitiser ions. The main reason, however, that some of them have been developed is that they have a fine lined multiple spectrum which allows them to be used in a tunable mode. For instance, alexandrite ($BeAl_2O_4 : Cr^{3+}$) can be lased over the whole of the 700 to 800 nm wavelength range and $Al_2O_3 : Ti^{3+}$ can be tuned from 675 to 945 nm. This is done by replacing the 100%R mirror with a prism and rotatable grating combination so as to allow only a narrow wavelength bandwidth to oscillate at any one time.

6.4 Electrically Driven Emission

Electrons can be raised from the ground state into the excited state levels by electronic means even more efficiently than by photonic means. This technology is used in electroluminescence, light-emitting diodes (LEDs) and semi-conductor lasers. As was explained earlier, there is very little difference in the physical principles between the three technologies, all relying on the electronic inversion and decay processes. The differences lie in the definition convention. In particular the differences between an LED and a laser diode are very minor (a laser diode operating as an LED below the oscillation threshold).

6.4.1 *Luminescence*

When a material absorbs energy a fraction may be re-emitted in the form of electromagnetic radiation in the visible or near infra-red regions of the spectrum. Luminescence is a process which involves at least two steps: the excitation of the electronic system and the subsequent emission of photons. Excitation may be achieved by bombardment with photons (photolumines-cence), with electrons (cathodoluminescence), as a result of chemical reaction (chemiluminescence), or by the application of an electric field (electroluminescence). The emission of photons consists of fluorescence (the emission of light during excitation) and phosphorescence (afterglow). The demarcation lines between these are not exact since there is always a time delay between excitation and the emission of a photon. Decay times of $^-10^{-8}$s are usually taken as the dividing line between the two phenomena.

The ability of a material to exhibit luminescence is usually associated with the presence of 'activators'. These may be impurity atoms occurring in relatively small concentrations in the host material or they may be small nonstoichiometric excesses of one of the constituents of the material.

The following classes of material exhibit luminescence:

(a)　Compounds which luminesce in the 'pure' state. These contain an ion or ion group per unit cell with an incompletely filled shell of electrons which is well screened from its surroundings (Randall 1939). Examples are the manganous halides, samarium and gadolinium sulfate, molybdates and platinocyanides.

(b)　The alkali halides activated with heavy metals such as thallium.

(c)　ZnS and CdS activated with Cu, Ag, Au, Mn or self activated with an excess of Zn or Cd. These are the most important electroluminescent materials in use at the present time and are used for large area flat panel displays.

(d)　Silicate phosphors, such as $Zn_2SiO_4 : Mn^{4+}$, which are used for oscilloscope screens.

(e)　Oxide phosphors such as ZnO and Al_2O_3 activated with transition metals.

(f)　Organic crystals, such as anthracene activated with naphtacene, which is used for scintillation counters.

Further recommended reading on this subject are articles by Randall (1939) and Garlick and Gibson (1948, 1949).

6.4.2　*Light Emitting Diodes*

Light emitting diodes (LEDs) are semiconductor junctions which emit light when an electrical current is passed through them. Their very low power consumption, long life, small size and high reliability has qualified them as indicator and display lights on a wide range of battery powered instruments.

Typical LED schematics are shown in Figures 6.8 (a, b). The device is fabricated as a matrix of devices on a GaAs substrate by either MOCVD (metalorganic chemical vapour deposition) or LPE (liquid phase epitaxy) and then cut up into pixels of $^-100\ \mu$m square devices. The active area of the pixel has dimensions of $^-10\ \mu$m by $50\ \mu$m and is usually encapsulated inside a resin mould either singly or, increasingly, as a quartet to yield a three-colour display. The wavelength of radiation is determined by the composition of the semi-conductor material used for the active layer and is therefore dependent on the precise fabrication conditions as well as by the starting materials. Binary (GaAs and SiC), ternary (GaAlAs) and quaternary (InGaAsP) semi-conducting compounds have all been used to fabricate LEDs and laser diodes. A plot of diode wavelength as a function of composition in the InGaAsP and GaAlAs systems is shown as Figure 6.9 (Hecht 1984, Kressel 1982). The range of wavelengths of operation available has gradually been extended, from the original red only, down to recent announcements of the successful operation of blue LEDs. At the present time a typical full colour LED display pixel consists of a red LED, a green LED and two blue LEDs (encapsulated inside the same diffusive resin moulds). This is because the blue LEDs have far less luminous intensity and the human eye is far more sensitive in the green. These LEDs are then used singly as indicator lamps or in a variety of increasingly complex geometries up to full screen large area arrays.

Photoluminescence in conjugated polymers results from radiative decay of singlet excitons confined to a single chain and is produced when light is

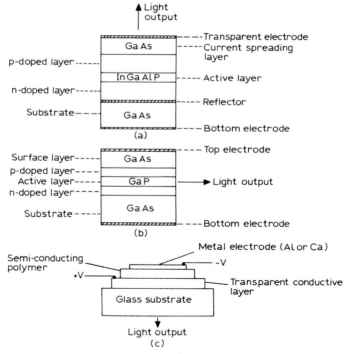

Figure 6.8   Light emitting diode schematics
(a) Single element of diode array
(b) Single chip LED
(c) Large area polymer LED

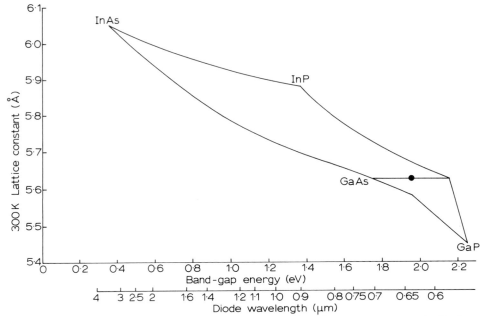

Figure 6.9   Plot of lattice constant versus band-gap energy and diode wavelength (from Kressel 1982)

absorbed across the semi-conductor band gap. Unless the polymer is very pure, however, the presence of dopants act as efficient quenching sites. However, using improved synthesis techniques photoluminescent efficiencies of 50% have been reached.

Figure 6.8(c) shows the extremely simple fabrication technology of a large area polymer electroluminescent device. When an electric field is applied across the polymer electrons and holes are injected at opposing electrodes and electron-hole capture forms bound excitons which can decay radiatively if they are in the singlet state. The negative, electron-injecting contact preferably has a low work function and the positive, hole-injecting contact has a high work function. One of the contact layers must be semi-transparent (e.g. indium tin oxide, ITO, or zinc oxide, ZnO_2). If a low work function electrode, e.g. metallic calcium, is used as the electron injecting contact layer then the operational voltage decreases and the photon per electron injected efficiency increases.

One of the main problems of polymer chemistry is that there are so many possible variations that it is possible to start from completely different structures to obtain approximately the same effect. Poly(phenylene vinylene) (i.e. PPV), poly(3-alkylthienene), and poly(phenylene) have all been used as the basis of efficient luminescent materials. The range of colours available is gradually widening as the chemists learn to engineer the polymers more exactly. For instance, PPV yields a wide band luminescence peaking in the yellow. A blue shift has been obtained by eliminating the methanol group and a red shift has been obtained by adding two alkoxy groups to the phenylene.

6.4.3 *Laser Diodes*

Semi-conductor laser diodes fall into two major categories:

(1) those fabricated from III–V compounds, e.g. GaAs, which emit in the near infra-red;
(2) those fabricated from lead salts which emit at longer wavelengths, e.g. 2.7 – 30 μm.

A diode laser is similar in structure to an LED with the addition of structural elements and reflective facets. A diode laser is a block of semi-conductor material containing a p–n junction. Current is carried by electrons, which are free to move within the crystal when an electric field is applied, and holes, which are vacancies for valence electrons within the crystal lattice. Unlike silicon diodes, where recombination of holes and electrons produces heat (phonons), III–V diodes generate a mixture of heat and light and hence a proportion of the recombination energy takes the form of a photon with energy equal to the band gap. This band gap is a function of the diode's crystal composition and working temperature and can therefore be tuned.

In LEDs light emission occurs spontaneously and is therefore incoherent. In a laser diode recombination of the excited carriers is stimulated by a photon of the recombination energy. In fact the resonator behaves just like a crystal laser (section 6.2), the resonator mirrors being produced by cleaving the opposing ends of the crystal to produce reflective facets to provide the optical feedback necessary to overcome the cavity losses and to sustain laser oscillation. As the stimulated emission process produces the strongest amplification at the peak of the gain curve it also suppresses emission at other wavelengths. The bandwidth of a diode laser is therefore much narrower than that of an equivalent LED. At low input currents most devices act as LEDs, at still higher currents the device behaves as an optical amplifier (see section 8.2), or superluminescent diode until the inversion is such as to permit laser oscillation to dominate.

There is an extensive range of diode laser structures and a representation of the main ones is shown in Figure 6.10 (Hecht 1984, Kressel 1982). The main trade-offs between the designs are the fabrication complexity (and hence the fabrication cost) and the output beam characteristics, such as divergence, spectral bandwidth, frequency stability and coherence length.

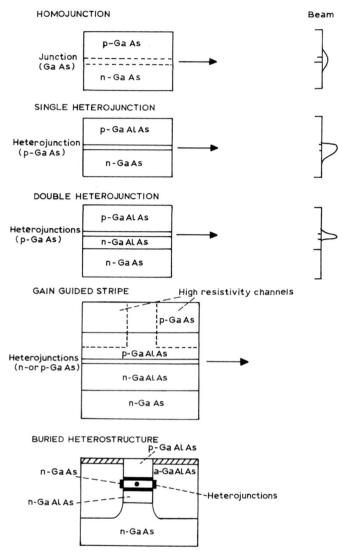

Figure 6.10   Diode laser structures

Possibly the most important market for laser diodes is the telecommunication industry which is vigorously competing in implementing the broad-band integrated services digital network (B–ISDN) of the future which will allow virtually unlimited possibilities in the realms of information technology. In order to achieve the maximum possible bit rates for this application layer designs have to be implemented which reduce the parasitic capacitance to a minimum. The favoured way of increasing the gain differences for stable, single frequency operation is to fabricate a corrugated structure of period $\lambda/2n$ inside the laser cavity (Lee 1989). The mode nearest the Bragg wavelength, λ, is reflected whilst the modes having higher losses are suppressed. Figure 6.11 shows (a) a cross section across a typical laser diode chip and (b) and (c) the schematic differences between a Fabry–Perot (FP) laser and a distributed feed-back (DFB) laser. Figures 6.11(d) and (e) show the pertinent spectral intensity curves, demonstrating that whilst the output of an FP laser can hop between modes the DFB structure is constrained to single mode operation.

For applications requiring a wide spectral range the laser designs need to incorporate tuning ability. This can be achieved by use of a cavity with an external grating, by use of a fibre extended cavity coupled to a fan-out grating or by use of monolithically integrated tunable designs. One such design takes advantage of the fact that the wavelength of a distributed feedback laser can be temperature tuned ($^-0.1$ nm/°C) because of the change of refractive index with temperature. In practice this is achieved by injecting current into the active Bragg region of the laser.

The main problems with obtaining high output power from laser diodes are that each diode has necessarily to be small ($10 \times 100\ \mu$m) in order to keep the beam divergence within tolerable bounds, and that as the output wavelength is temperature sensitive it is not possible to run two-dimensional arrays fabricated on a single substrate at high input powers. In order to overcome this problem a bar-in-groves approach has been shown to yield good results.

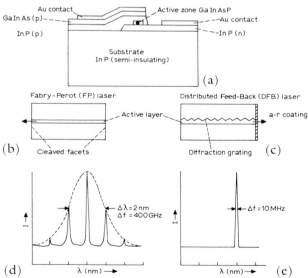

Figure 6.11   Laser diode schematics and spectral characteristics

By using highly conductive materials (such as boron nitride or diamond) as heat sinks it has been demonstrated that peak optical pulse densities of ˜40Wmm^{-2} can be obtained from a 10×10 mm array module.

Two novel laser designs are the etched microdisc approach pioneered by Bell Laboratories, (Murray Hill, USA) (Hecht 1991) and columnar lasers developed by RSRE (Malvern UK) (Cullis and Canham 1991). The first consists of 10 nm quantum wells of InGaAs sandwiched between 10 nm barrier layers of InGaAsP. The material is grown by electron-beam epitaxy and etched leaving drawing pin-like structures standing proud of the InP substrate. These lasers operate differently from a conventional laser since instead of oscillating between a pair of mirrors the light circulates round the rim of the disc, confined by total internal reflection (because $n = 3.5$ for the semi-conductor and $n = 1$ for air). Laser light ultimately emerges from the side and the top of the discs. In practice a single diode would comprise an array of these structures. In order to make the structures more robust they can be encapsulated in SiO_2 (which still leaves an appreciable r.i. difference). The RSRE approach has been to etch thin columnar vertical cavity lasers (˜0.5 μm in diameter and 10 μm height). These have been used to generate a whole spectrum of laser wavelengths, from the red down to the blue, and may well be used as the source for full colour displays in the future. Both these approaches are radically different from that which has gone before because conventional theory says that since these semi-conductor materials have relatively small indirect band gaps their radiative efficiency should be very inefficient and that virtually all the excited carriers should recombine non-radiatively to produce heat and not light. The effect becomes possible only because the bandgap increases from ˜1eV to up to 7 eV as the dimensions of the material are reduced to the 1–5 nm size regime. It will therefore be realised that for both 'quantum wire' (2-D confinement) and 'quantum dot' (3-D confinement) the bandgap is increased and, in principle, emission over the whole of the visible spectral range should be possible.

CHAPTER SEVEN

Detector Materials

7.1 Introduction

An optical detection system intercepts part of an incident light beam and converts it into some other sort of energy (photonic, thermal, acoustic or chemical). In most cases the ultimate output signal is electronic. There are a variety of optical detectors broadly falling into three categories, viz. quantum, thermal and photoemissive. Within each category there are a range of differences which principally show up in the detector characteristic, in particular the wavelength, sensitivity, linearity and speed.

The most common application for each type of photodetector is listed in Table 7.1. This table is not exclusive and there is no clear cut divide between the photodetector type and the usage. Nevertheless the table does give a guide to normal usage of these devices.

The following sections discuss the material parameters, device operation and design and the operating characteristics for both quantum detectors and thermal detectors. A short section describing the characteristics of photo-multipliers and vacuum photodiodes is also included for completeness and comparison.

7.2 Semi-Conductor Detector Materials

The simplest form of semi-conductor detector is the photodiode. This is an optical detector fabricated from semi-conductor material within which there exists a region which is depleted of free carriers (due to either an internal or an applied electric field). When a photon is absorbed within this region it generates an electron-hole pair which is then swept out by the field to give rise to a current in an external electrical circuit. The simplest form of this is the PiN photodiode which consists of a layer of intrinsic semi-conductor material sandwiched between a p-type doped layer and an n-type doped layer (see Figure 7.1).

Semi-conducting detectors can be used in either the photoconductive or the photovoltaic mode. Basically, when photons are absorbed electrons are

Table 7.1   Detector types and applications

Detector type		Wavelength (μm)	Application	Comments
Si photodiodes		0.2 1.1	General detection	Fast, linear sensitive
Si	APD		Laser range finding	Very fast
Si quadrant det.			Metrology	Balanced quadratures
Si arrays	CIDs CCDs		Atomic spectroscopy 2-D cameras Video camcorders Microscope recorders	Fewer problems with overload and dark currents than PMTs
Si based	PMTs		Low light applications	Ultrasensitive Very fast
Ge		1 2	Near IR detection	Non-linear characteristics Very temperature sensitive
InGaAs PtSi PbS InSb		1 2.5	Fibre optic data and telecommunications Lidar, receivers Imaging Surveillance, astronomy	
HgCdTe			IR laser range finders Target designators Heat seeking missiles IR and Fourier Transform Spectroscopy Thermal imaging	Mainly developed for military equipment Scientific instrumentation Fire fighting
HgMnTe			Laser range finders	High speed avalanche photodiodes
Pyroelectric	LiTaO$_3$ TDS PbS		Thermal sensors Intrusion alarms	Wide band

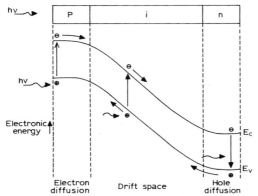

Figure 7.1   PiN photodiode schematic

excited either from the ground state into intermediate excited state holes or from these excited states into the conduction band. In the photoconductive mode (forward bias) photons produce additional electron-hole pairs which alter the electrical conductivity and hence give a change of current. Charge-coupled devices, CCDs, and charge-injection devices are examples of this effect. If, on the other hand, the semi-conducting junction is reverse biased the photovoltaic effect applies. In this mode highly doped *p-n* junctions are reverse biased, the incident photons are absorbed at the *p-n* junction thus increasing the number of free electron-hole pairs, causing a change in voltage which can be measured by an electrical circuit. This mode is commonly used for low level pulse detection (e.g. avalanche photodiodes). In general diodes operating in the photovoltaic mode are faster and more linear but more subject to output saturation than those operating in the photoconductive mode.

The useful range of detection sensitivity is determined by the band gap of the semi-conducting material, and hence by the absorption spectrum of the material. There are a range of semi-conducting materials, as can be seen by perusal of Table 7.2. This table lists the detector characteristics of the main materials. Figure 7.2 shows the spectral range and Figures 7.3, 7.4, 7.5 and 7.6 the wavelength/detectivity curves for these materials. It will be noticed that all the wavelength/detectivity curves are much the same shape which is set by the absorption band associated with the natural frequency of the lightest atoms of the material (see section 2.4). It will also be noticed that the spectral differences are in line with this, the heavier atoms being associated with absorption and detectivity at longer wavelengths than the lighter atoms. This fact has been used to advantage in the tailoring of the sensitivity curves for detectors in the far infra-red (e.g. notice the difference in the spectral sensitivity curves for small changes in the compositional make-up of mercury cadmium telluride, HgCdTe). It will furthermore be noticed that the

Table 7.2   Characteristics of quantum detectors

Detector type		Spectral range (μm)	Responsivity	NEP $W/\sqrt{H_z}$	Specific detectivity cm $\sqrt{H_z}/W$	Operating temperature
Silicon	PiN	$0.2 \to 1.2$	0.6 A/W	10^{-16}	7×10^3 at 1 μm	$-40 \to +80°C$
Silicon	PV	$0.18 \to 1.15$	$0.05 \to 0.6$ A/W	10^{-15}	–	Ambient
Silicon	PC	$0.18 \to 1.15$	$0.1 \to 0.6$ A/W	10^{-14}	–	$-50 \to +80°C$
Silicon	APD	$0.4 \to 1.1$	to 120 A/W	10^{-15}	–	$-50 \to +80°C$
Germanium	PiN	$0.5 \to 1.8$	0.9 A/W	10^{-13}	3×10^{12} at 1.4 μm	$-30 \to +75°C$
Germanium	PC	$0.5 \to 1.8$	$0.15 \to 1$ A/W	10^{-15}	3×10^{12} at 1.4 μm	$77K \to$ ambient
Germanium	APD	$0.8 \to 1.8$	0.2 A/W	10^{-10}	–	–
InGaAs	P	$0.9 \to 1.65$	0.6 A/W	10^{-12}	10^{11} at 1.3 μm	Ambient
InGaAs	APD	$0.9 \to 1.7$	0.8 A/W	10^{-13}	10^{11} at 5 μm	$-40 \to +80°C$
InSb	–	$1.5 \to 5$	$1 \to 3$ A/W	10^{-17}	10^{10} at 3.4 μm	77K
InAs	–	$1 \to 3.8$	$0.8 \to 1.4$ A/W	10^{-14}	–	$77K \to$ ambient
InGaAs	PiN	$1 \to 1.7$	$0.6 \to 0.7$ A/W	10^{-4}	–	$-40 \to +80°C$
PbSe	Ambient	$1 \to 4.8$	$2 \to 6$ kV/W	10^{-11}	10^9 at 4.5 μm	Ambient
PbSe	TE-cooled	$1 \to 5.5$	$5 \to 10^4$ V/W	10^{-11}	10^{10} at 4.8 μm	193K
PbSe	LN$_2$ cooled	$1 \to 6.8$	$5 \to 10^4$ V/W	10^{-12}	10^{10} at 6.5 μm	77K
PbS	Ambient	$1 \to 3$	$0.1 \to 2$ A/W	10^{-11}	10^9 at 2.7 μm	Ambient
PbS	TE-cooled	$1 \to 3.3$	10 A/W	10^{-12}	10^{10} at 3.0 μm	193K
PbS	LN$_2$ cooled	$1 \to 4$	10 A/W	10^{-13}	10^{11} at 3.7 μm	77K
HgCdTe	PC	$2 \to 20$	$\to 5$ A/W	10^{-12}	10^{10} at 10 μm	$Ambient \to 77K$
HgCdTe	PE/Mag	$3 \to 12$	$\to 0.01$ V/W	10^{-7}	10^6	Ambient
HgCdTe	PV	$8 \to 12$	5 A/W	10^{-16}	3×10^{11} at 10.5 μm	77K
Silicon, Ga doped		$8 \to 17$	2 A/W	10^{-16}	–	4K
Germanium, Ga doped		$60 \to 120$	8 A/W	10^{-17}	–	3K
Germanium photon-drag		$9 \to 11$	10^{-2} V/W	10^{-3}	–	Ambient

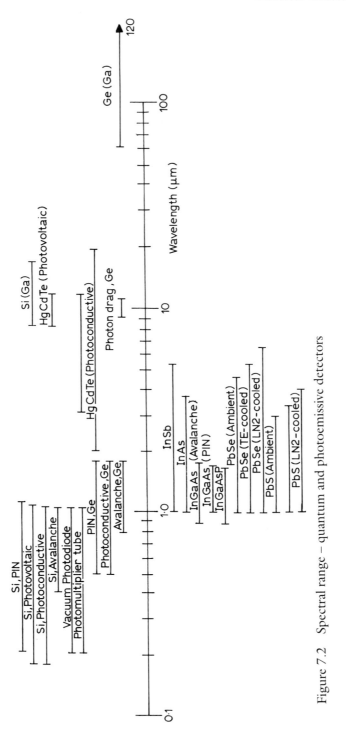

Figure 7.2 Spectral range – quantum and photoemissive detectors

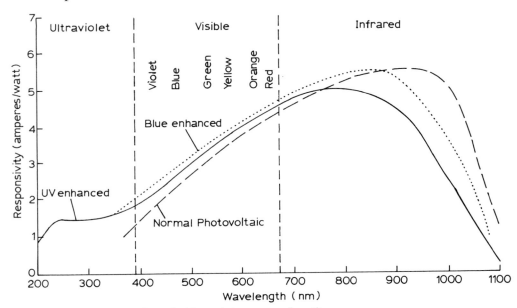

Figure 7.3   Wavelength/detectivity curves for silicon

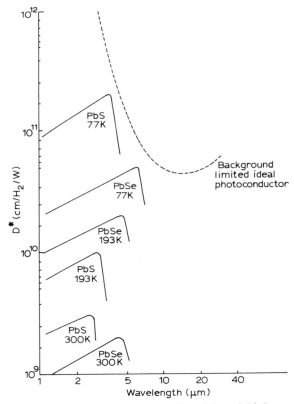

Figure 7.4   Wavelength/detectivity curves for PbS and PbSe

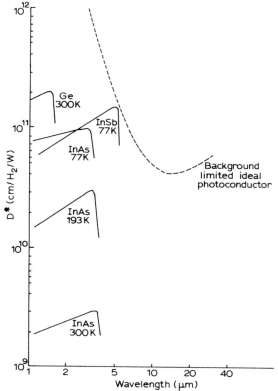

Figure 7.5   Wavelength/detectivity curves for Ge, InAs, InSb, InGaAs

detectivity increases with increasing temperatures and that the peak of the spectral sensitivity curve also moves (to a longer wavelength) with temperature. This behaviour can all be described from basic physical principles as the electron orbits move and the effective band gaps change slightly. The main points that must be taken into account when specifying the use of a quantum detector are that both the sensitivity and the dark current are temperature dependent and that at high optical fluxes there comes a point where the sensitivity curve becomes markedly non-linear. In most cases this threshold marks the limit of useful operation and continued operation above this input level can lead to melting of the semi-conductor junction and ultimately catastrophic damage.

The characteristics (except for the basic wavelength range) of a photo-conductor are controlled by the precise fabrication details and therefore the responsivities, noise equivalent powers, non-linear threshold and speeds vary from type to type. The thickness of the intrinsic layer is the major factor in the performance of a particular device. For high speed operation the *i*-doped layer

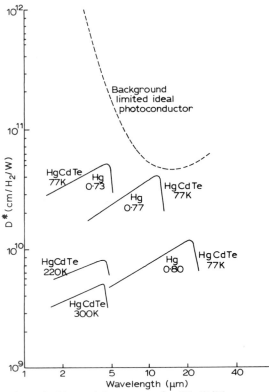

Figure 7.6   Wavelength/detectivity curves for HgCdTe

needs to be thin in order to reduce the time taken for the photogenerated carriers to traverse it. This, however, has adverse effects – a reduction in the number of photons absorbed resulting in a low sensitivity device and ultimately a reduction in the speed of response because of an increase in the RC time constant. Diffusion of the photogenerated minority carriers outside the depletion region can occur and this needs to be minimised for high speed operation. One solution is to make the device small so that it reduces both the junction and the parasitic capacitance. Another solution is to reduce the p-layer thickness to allow the light to pass into the i-layer. Unfortunately this increases the series resistance of the diode which again degrades the speed of the device. There are a number of imaginative solutions which have been used in the design stage to overcome these problems (Parker 1990). Speeds of up to 110 GHz, responsivities of >0.9 A/W and low leakage currents of 100 pA have all been demonstrated, albeit not all at the same time. In short, the precise characteristics required from each photodiode need to be considered carefully in the light of the application in order to gain the maximum advantage.

The operation of an avalanche photodiode (APD) is the same as for the PIN photodiode except that it has internal gain. The avalanche photodiode is operated near to its breakdown voltage and once an electron-hole pair is generated avalanche occurs. This is usually used in applications where timing (fast response) is required without the requirement for linearity of response. A current limit has to be applied to stop the diode damaging catastrophically under these operating conditions.

Quadrant detectors are specialised arrays with four arc sections and are mainly used for metrology, the operation relying on a well collimated, uniform cross section laser beam for accurate location and/or angle sensitivity.

Arrays of individual photodiodes can be fabricated on a single semiconductor chip. These can be operated either individually or more commonly using scanning addressing techniques. These techniques allow closer packing to be implemented. Two common technologies are charge injection, CID, and charge-coupled, CCD, devices. The CID array injects charge into a sense node and the CCD shifts the signal charge in series to an output sensing node (Messenger 1992). These devices do not experience any of the adverse sensitivities that ordinary photodiodes or photomultipliers possess and can be exposed to high irradiance without fear of burnout or high noise levels. They also exhibit virtually zero dark current when cooled and are thus not limited by dark current shot noise (Dereniak and Crowe 1984).

Signal-to-noise ratios of array detector systems are less than for single element detection systems (by $N^{1/2}$ where N is the number of detector elements). In low light level cases the ratio is even better because when detector noise limits the measurement arrays have an advantage of N because of the square root dependence on time of the noise in PMTs and photoconductive detectors while the noise in charge coupled detector arrays is associated with a single read-out of the integrated charge. As arrays are currently being fabricated compatible with standard TV definition the advantage is considerable.

Solar cells are constructed from large area thin blanks of polycrystalline silicon and are arranged so that they lie in a large area array facing the sunlight. They can be operated in either the photoconductive or the photovoltaic mode. The current emitted is either used to warm up coolant flowed gently past the array or to charge up batteries for future use. Applications range from heating (a common sight in Mediterranean and equatorial countries) to satellite power and even to large-scale electrical generation (e.g. on the West coast of America).

One of the most common detectors used in the far infra-red with CO_2 lasers is the photon-drag detector. This is fabricated from a block of germanium with longitudinal ring electrodes and operated at a wavelength where the absorption is significant but not total (see Figure 7.7). The detector can thus be used as a transmissive in-line monitor and is frequently used to

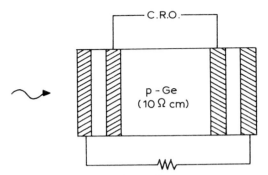

Figure 7.7   Photon-drag detector schematic

monitor the pulse shape of pulsed CO_2 transverse-excited atmospheric pressure (TEA) lasers. The input pulse stimulates electrons from the input end of the device which are then swept towards the output end and the charge is collected by the rear ring electrode. The device is neither very linear or sensitive but has the distinction of being the only transmissive, fast, real-time monitor available in the far infra-red. The design reproduced in Figure 7.7 (Edwards *et al.* 1983) contains a second pair of electrodes which helps keep the spatial characteristic flat.

7.3   Photoemissive Detectors

Two very important types of detector are the photomultiplier and the vacuum photodiode. These are used extensively in the visible and infra-red regions of the spectrum and although they can have large detection areas, and are therefore very sensitive, they are also very fast (Edwards and Jefferies 1964). Their main drawback is that they both need high voltages to run them at maximum efficiency and that they are relatively bulky. The spectral ranges of both types of detectors are included in Figure 7.2 and the main characteristics are tabulated in Table 7.3. The details of the operation of these devices is outside the scope of this book although it has been thought fit to include the characteristics for comparison with those of the quantum and thermal detectors.

7.4   Thermal Detectors

Thermal detectors basically absorb the light and operate a detection mechanism based on the measurement of temperature. The main types are tabulated in Table 7.4, and include bolometers, calorimeters, thermopiles and

Table 7.3 Characteristics of photoemissive detectors

Detector type	Spectral range (μm)	Responsivity	NEP $W/\sqrt{H_z}$	Specific detectivity cm $\sqrt{H_z}/W$	Operating temperature
Photomultiplier	$0.2 \to 1.1$	0.9 A/W	–	–	$-50 \to +50°C$
Vacuum photodiode	$0.2 \to 1.1$	0.001 A/W	–	–	Ambient

Table 7.4 Characteristics of thermal detectors

Detector type		Spectral range (μm)	Responsivity	NEP $W/\sqrt{H_z}$	Specific detectivity cm $\sqrt{H_z}/W$	Operating temperature
Bolometer	Si	$1.6 \to 5000$	$10^5 \to 10^8$ V/W	10^{-16}	–	$\to 4K$
Bolometer	InSb	$250 \to 6000$	10^3 V/W	10^{-3}	–	$\to 4K$
Bolometer	Thermister	$0.25 \to 500$	300 V/W	–	10^8 at 2.2 μm	$-55 \to +80°C$
Calorimeter		$0.25 \to 35$	0.1 V/W	10_w^2cm^{-2}	–	Ambient
Pyroelectric	LiTaO$_3$	$0.01 \to 1000$	1μ A/W	10^{-10}	10^8 flat	Ambient
Pyroelectric	SBN	$0.1 \to 1000$	0.25μ A/W	–	–	Ambient
Pyroelectric	DTGS	$0.25 \to 500$	10^3 V/W	10^{-10}	10^9 at 2.2 μm	Ambient
Thermopile		$0.2 \to 35$	50 V/W	–	10^6 flat	Ambient

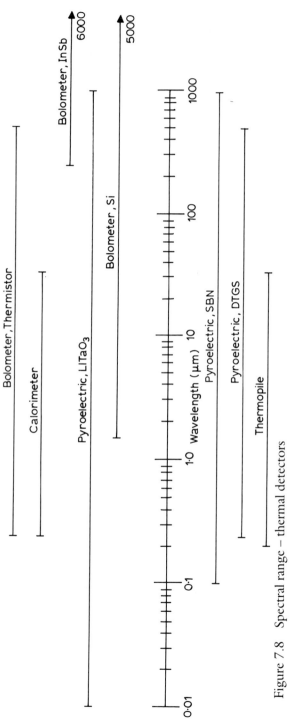

Figure 7.8   Spectral range – thermal detectors

pyroelectric detectors. The spectral ranges of these detectors are shown in Figure 7.8 and can be seen to extend over a very wide band. Most of the spectral characteristics are flat over the whole range, any variations being due to the differences in the reflection characteristics of the surfaces.

Pyroelectric detectors are fabricated by placing a thin layer of pyroelectric material, such as lithium tantalate, between a heat sink and a thin absorbing layer. The absorber converts the incident radiation into heat which is then transferred to the pyroelectric layer. This results in a transient voltage output, the peak of which is proportional to the total pulse energy. The rise time of the voltage output is determined by the time constant of the absorber, and the decay time by the RC time constant of the electrical circuit. The device is usually used to integrate pulse energy (RC – usually μs to ms); however, detectors with nanosecond RCs have been made (at the cost of sensitivity) to give real-time displays of 100 ns FWHM laser pulses.

The most accurate, robust and linear photonic detectors for the measurement of integrated laser pulse energy, and which are therefore invariably used for standards purposes, are calorimeters. The calorimeter simply consists of a piece of absorbing material, which absorbs the incident radiation, and a series of thermocouples or thermistors which convert the heat into an electrical signal. The absorbing material is usually in the form of a disc or a cone. Discs can be black painted metal or semi-absorbing transparent material (e.g. glass for use in the visible) and cones are usually made of either black painted metal foil or carbon. Useful references regarding the fabrication, calibration and use of calorimeters may be found in articles by Edwards (1967, 1970) and Gibbs and Lewis (1978).

It will be noticed from comparison between Tables 7.2 and 7.4 that while the responsivity of most quantum detectors is quoted in A/W, that of most thermal detectors is quoted in V/W and that the noise equivalent powers (NEP) and specific detectivities are lower for thermal detectors than for quantum detectors. In addition, although thermal detectors can be used to monitor fast events they usually have to be run at low detectivity and/or can only respond to fast increases and not decreases in photon flux. It is instructive to note that although the operating characteristics of thermal detectors are rarely as good as the equivalent quantum detectors they are much more stable (having less drift with life, having flatter spectral responses and being less temperature sensitive) and are usually used as the basis of 'standard' systems where these characteristics are valued more highly than absolute detectivity.

CHAPTER EIGHT

Fibre/Integrated Optics

8.1 Introduction

One of the radical revolutions which swept the world in the twentieth century was the advent of long distance communications – radio, telephone, television. These means of communication have shrunk the world beyond recognition, revolutionising politics, business and leisure. A second revolution has also been going on within the new-born communication industry, quietly replacing most of the old copper wire-based technology used in these systems by optical fibres and optical waveguides.

The largest usage of optical fibre technology at present is in the area of optical fibre trunk cables. These have now superseded the older copper wire technology in the industrialised countries of the world (Midwinter and Guo 1992). The key requirements for such land-based links are high capacity and low cost per channel. The next phases of the revolution will be to replace the older technology in the areas of intercontinental communication and in the local information-carrying networks that serve the individual subscriber or user. The first usage demands ever longer distances between repeater links and the second the development of integrated optic circuitry to sustain the advantages of parallel all-optical processing over standard serial electronic processing.

The particular advantages that lightwave technology has over the alternative radiowave technology, used to advantage in satellite communications, are that it is 'secure' and interference free and has frequencies more than five orders of magnitude higher. These advantages outweigh the necessity of providing optical fibre links for many of the applications. It must of course be emphasised that optical technology can be used in free space environments in the same way as radio or radar (i.e. laser rangefinding, lidar etc. – see Table 7.1 for a list of the major applications of laser sources). However, although free space optical communication is non-interfering it is directional and depends on line of sight and/or reflectors which are not acceptable for the mainstream telecommunication industry. A useful description of the whole field of opto-electronics and lightwave technology is given by Midwinter (1992).

If optical fibres are to be used to their greatest advantage then it will be necessary to develop all-optical circuitry to perform such functions as pulse amplification, beam modulation, analogue-to-digital conversion and optical switching without converting the signals from optical to electrical and back again. This is necessary not only for reasons of compactness of the system but also to allow the inherent speed of the optical frequencies to be used to maximum advantage.

The concept of integrated optics emerged in 1969. This was first envisaged as being miniature optical components butt-jointed together. This approach raised problems, particularly of alignment and the use of unattractively high voltages. It soon became apparent that a completely monolithic approach was preferable and subsequent development has led to the whole process being confined to the first $10 \, \mu m$ or so of the surface of suitable substrates, the substrate only being there to provide strength and alignment to the system. Static optical devices such as beam splitters, couplers and wavelength demodulators are fabricated in or on the surfaces of silicon, fused silica or alumina substrates, whilst active devices such as amplitude modulators, switches and interferometers are fabricated in $LiNbO_3$ or GaAs. This technology is now available in high optical quality devices. The waveguide-based technology has, interestingly enough, proved to be even more compatible with butt jointing or holographic interconnects to the main optical fibre link than the discrete component approach.

8.2   Fibre Optics

The heart of a single-mode fibre optic cable is comprised of thin glass fibres each of the order of $10 \, \mu m$ diameter, clad with a $125 \, \mu m$ diameter lower index glass to ensure that any optical ray launched into the end of the fibre will travel down the fibre with minimal loss because of total internal reflection of the beam trapped in the core (see Figure 8.1a). This figure introduces the concept of a numerical aperture (NA) which is a measurement of the angle at which the rays escape from the core into the cladding:

$$NA = \sin \theta = / \, n_1^2 - n_2^2$$

where n_1 is the refractive index of the core material
n_2 is the refractive index of the cladding

The reasons why optical fibre is becoming increasingly preferred over the older and more conventional copper wire technology stem from consideration of the relative cost, size, information-carrying capacity and speed of the two technologies. The attenuation of optical fibre is much better than copper cable (being ¯0.2 dB km^{-1}, the optical power being halved every 15 km). In

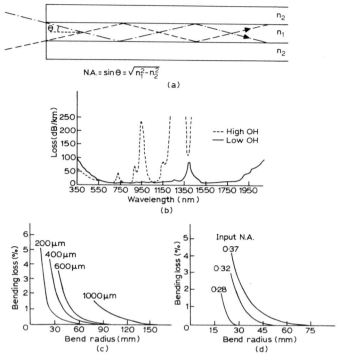

$$\text{N.A.} = \sin\theta = \sqrt{n_1^2 - n_2^2}$$

(a)

(b)

(c)

(d)

Figure 8.1   Optical fibre characteristics
(a) Definition of numerical aperture, NA
(b) Spectral transmission curves for high and low –OH Fibres
(c) Bending loss versus bend radius and core diameter
(d) Bending loss versus bend radius and launching angles

addition, standard copper cable has an attenuation which rises as $v^{\frac{1}{2}}$, the frequency of the modulation, whereas glass fibre offers attenuation coefficients which are constant with signal bandwidth up the >50 THz. A third factor is that electromagnetic interference is absent in fibres whilst necessitating stringent precautions in copper cable. In addition to these almost overwhelming advantages are the facts that, because of the much greater bandwidth and information-carrying capacity and as the basic material, SiO_2, is so abundant and as the diameter of the fibre cable is so much smaller, the manufacturing and installation costs make the change to optical fibre technology almost inevitable.

Most of the land-based optical fibre cable systems at present in use work with silica-based fibre in the wavelength range 1.3 – 1.5 μm where the fibre attenuation is lowest (see Figure 8.1b). The attenuation value of 0.2 dB km^{-1} is well quantified. Silica fibre is available both as single-mode fibre (~10 μm diameter in a 125 μm cladding) and multi-mode fibre (up to 1 mm diameter fibre is readily available). Single-mode fibre is used for high bit rate communication purposes and multi-mode fibre is used for display and lighting purposes where the beam does not carry significant amounts of information. It should be realised that multi-mode fibre is much more sensitive to fibre bending losses than single-mode due to the larger angles subtended by the higher order modes. Figure 8.1(c) shows the bending loss versus bend radius as a function of the core diameter for 0.37 NA silica fibre, whilst Figure 8.1(d) shows the bending loss versus bend radius as a function of the input numerical aperture for 400 µm diameter silica fibre. A very good article on this subject is by Boechat *et al.* (1991). The use of fluoride-based glass was seriously considered for long haul communications during the 1980s because the theoretical attenuation of this material has been calculated to be ~0.01 dB km^{-1} at ~2.5 µm wavelength. Moving to this wavelength region would necessitate development of both 2.5 µm laser sources and detectors as well as the fluoride fibre itself. This would tend to delay the transfer of the technology because economic considerations mitigate against developing a new technology, even one with proven advantages, if the older one can do the job adequately. However, due to problems of materials development the attenuation of fluoride fibres is still dominated by submicron scattering and the theoretical loss is unlikely to be achieved without a vapour deposition route.

Problems still exist even for the extremely low loss attenuation fibre at present available. The main problems are that both attenuation and dispersion of the optical pulse width occur as the pulse travels down the fibre. These attenuate and broaden the pulse and result in degraded readability of the data stream. Repeaters are therefore necessary both to amplify and to reshape the optical pulse. It must also be noted that the input pulse power cannot be raised indefinitely because Brillouin scattering, Raman scattering (see Chapter 5), self-focusing (see Chapter 3) and ultimately laser-induced damage (see Chapter 10) all have relatively low power thresholds in the small cross sectional areas involved.

Early repeaters consisted of photodiode detectors combined with electronic shaping circuits and laser diode sources. More recent ones are all-optical and may consist of semi-conductor amplifiers or fibre laser amplifiers. Both types of amplifier are based on conventional laser amplifier designs, the resonator mirrors being removed from a laser source and the signal amplified in a single pass configuration. At present doped fibre (e.g. Er^{3+} : silica) amplifiers have a better performance than the semi-conductor amplifiers due to symmetry considerations and due to the fact that doped fibre can be spliced directly to the main fibre with minimal loss, low reflectance at the joint and minimal extraneous 'ripple' induced on the signal. Both types of amplifier require the provision of either electrical or optical pump energy. This is usually done using a fibre coupler which combines the pump and the signal, each in a separate fibre, into a single output fibre.

Unlike electronic repeaters, optical amplifiers cannot fully reconstruct the original shape of the optical pulse and therefore the pulse width ultimately degrades and adversely affects the information-carrying capability of the fibre. Fortunately a special sort of optical pulse, termed a soliton, has been discovered which gets over this problem. A soliton has particular shape and intensity characteristics so that the two major sources of temporal deformation – chromatic dispersion and non-linear refraction – work against each other and cancel each other out. It is also significant that in the silica fibres used for optical transmission this compensation occurs around the 1.5 μm attenuation minimum. Soliton pulses can therefore propagate in optical fibres over thousands of kilometres and still keep their shape, as long as they receive a periodic amplification boost to maintain their average energy. A good treatment of this subject is given by Taylor (1992).

8.3   Passive Optical Waveguides

Early work in the subject concentrated on the provision of such components as passive couplers (for distributing optical signals equally from a single fibre into a number of fibres), switches (changing optical paths of a pulse train by injecting a signal into another fibre) and wavelength division multiplexers and demultiplexers (to introduce or to separate different frequencies in a fibre system).

These components are fabricated using chemical vapour deposition technology (SiO_2 deposition cladding round a GeO_2/SiO_2 substrate) (Nourshargh *et al.* 1985, 1986), proton exchange (P_2O_5 core in an SiO_2 matrix) (Najafi 1992) or ion diffusion (Ti waveguide on a $LiNbO_3$ substrate). The first two technologies are equally good and have the advantage of leaving a completely clad waveguide. Simplified details showing the fabrication technology are given in Figures 8.2 and 8.3. Packaged devices fabricated using both technologies are available and have excellent characteristics. The main advantage of the CVD technology is that it is totally compatible with the fibre technology. A selection of the passive devices available are shown in schematic form in Figure 8.4. The possible devices include couplers, splitters and wavelength division multiplexers and demultiplexers. Figure 8.4(a) illustrates the schematic for a 1 : 8 coupler. As long as all the splitting angles are smooth and equal this arrangement ensures that the single input is divided into eight very accurately and with minimum loss. Splitters with ratios up to 1 : 32 and star couplers with ratios up to 32 : 32 have been fabricated successfully. Figure 8.4(b) illustrates the schematic of a directional coupler (no power being

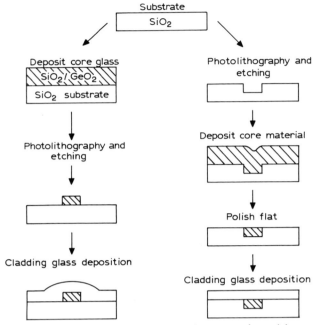

Figure 8.2 Waveguide fabrication – chemical vapour deposition

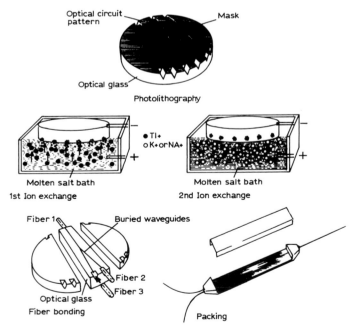

Figure 8.3 Waveguide fabrication – proton exchange

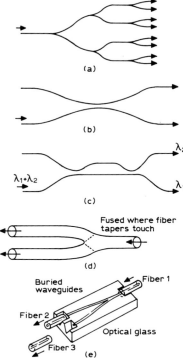

λ_2

$\lambda_1 \cdot \lambda_2$

λ_1

Fused where fiber tapers touch

(d)

Buried waveguides

Fiber 1

Fiber 2

Optical glass

Fiber 3

(e)

Figure 8.4   Passive waveguide devices
(a) 1:8 coupler schematic
(b) Directional coupler schematic
(c) Passive wavelength division multiplexer
(d) Fused biconical taper coupler
(e) Butt joint in grooves, fibre coupler schematic

reflected up the fourth arm whilst still allowing the possibility of an input from that direction). Figure 8.4(c) illustrates the operation of a passive wavelength division multiplexer/demultiplexer (depending on the direction of the input). The performance of these devices is well understood and the wavelength/ frequency isolation is very high. The main problems lie more in the prevention of strain at the integrated device/fibre interface rather than the efficiency of the join itself. For this reason alone it is more sensible to make a completely integrated device than to rely on separate, linked components. Future work will inevitably be on the integration of these devices and the provision of whole optical systems. Fibre fusion splicing and v-groove/epoxy interconnections (see Figures 8.4(d) and (e)) have both been shown to be very efficient and fibre/fibre couplers are now available on the open market.

8.4   Active Optical Waveguides

Chapter 5 included a description of such phenomena as stimulated Raman and Brillouin scattering, two photon absorption, harmonic generation, up-conversion and parametric oscillation. All these processes are a function of the non-linear coefficients of the materials, the power confinement (power density times the interaction length), the phase matching and the overlap of the interacting waves. Integrated optical waveguides allow these effects to be implemented at relatively low input powers because the guided wave configuration allows the optical fields to be concentrated into long structures with cross sections of the order of a few wavelengths (deMicheli and Ostrowsky 1990).

This is illustrated by reference to Figure 8.5 which compares the bulk focal and the integrated optical configuration. The efficiency of the SHG conversion is given by:

$$\text{eff} \approx \frac{d^2 L^2 P_\omega}{s} \sin^2\left(\frac{\Delta k L}{2}\right) I_r$$

where d is the non-linear coefficient
   L is the interaction length
   P_ω is the fundamental optical power
   s is the beam cross section

(i.e. $L^2 P_\omega / s =$ power density confinement/interaction length
   I_r is the overlap between the fundamental and the harmonic waves

   $\Delta k = k_{2\omega} - k_\omega$

   k_ω, $k_{2\omega}$ – the wave vectors at frequencies ω and 2ω
   $\sin^2 (\Delta k L/2)$ describes the effect of optical phase matching between the fundamental and the harmonic waves

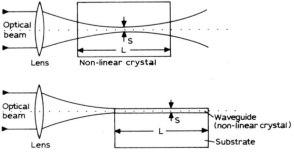

Figure 8.5   Comparison between confocal bulk and integrated optics configurations

In the bulk case diffraction leads to a compromise between a small value of the beam cross section, s, (and hence a high power density), and a large value of the interaction length, L.

For Gaussian beams the optimum configuration is confocal focusing, where

$$s = \lambda L/2n$$

where n is the refractive index of the waveguide material

In the waveguide case all the energy is focused inside the high refractive index region, cross sectional area s, and the ratio of the guided wave structure power density to that of the confocal configuration is then $\lambda L/2\ ns$. For the practical case where $n = 2$, $\lambda = 1\ \mu m$, $L = 10\ mm$ and $s = 25\ \mu m^2$ the improvement is $\times 1000$. This can of course be improved further by using an even longer guided wave section. This becomes increasingly more expensive in the bulk case because of the component volume necessary.

Guided wave optics can allow phase matching ($\Delta k = 0$) via modal dispersion in cases where the natural crystal birefringence does not and allows the waveguide to be fabricated from materials which cannot be used in the bulk configuration. The efficiency of the guided wave interaction may be reduced by the overlap integral I_r. As both the fundamental and the harmonic beams in the bulk are Gaussian, the overlap is good. In the case of guided wave interactions the overlap factor varies as a function of the interacting modes and care must be taken to design the component with reference to the wavelengths of interest.

In practice most active, non-linear waveguides have been fabricated using LiNbO$_3$. Two techniques have been used to obtain good waveguide properties: titanium indiffusion (TI) and proton exchange (PE). In TI a Ti layer ($^\sim$500 Å thick) is diffused into the surface of the LiNbO^3 over several hours at $^\sim$1200°C. In PE a crystal is immersed in acid at $^\sim$200°C allowing Li to be exchanged with protons. A useful article explaining the process is given by deMicheli and Ostrowsky (1990). It is expected that more use will be made of GaAs as a waveguide substrate in the future as this will aid fuller integration.

Figure 8.6 illustrates the operation of three different active waveguide devices. Figure 8.6(a) shows the schematic of a simple phase modulator. This consists of a metal electrode covering part of an active channel (AlGaAs strip on a GaAs substrate). When a voltage is applied to the electrode the phase of the light being transmitted along the waveguide is changed by an angle proportional to the voltage. Figure 8.6(b) shows how this phase modulation can be utilised when combined in an optical Mach Zhender interferometer. Constructive and destructive interference occurs between the light beams traversing the two channels and the transmitted light beam is modulated in intensity. This can be used to switch as well as to modulate the beam and

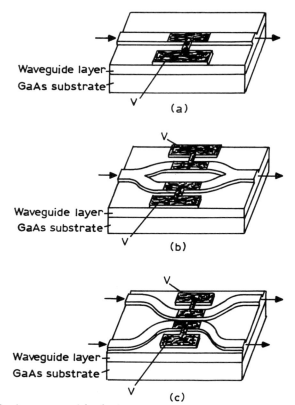

Figure 8.6   Active waveguide devices
(a) Phase modulator
(b) Mach zehender intensity modulator
(c) 2 × 2 directional coupler

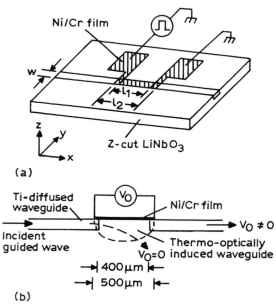

Figure 8.7   Thermo-optical waveguide switch

Figure 8.6(c) shows the schematic of a 2×2 directional coupler working on the same principles.

Active waveguide devices can be fabricated using the acousto-optic and thermo-optic as well as the electro-optic effects. Figure 8.7 shows the construction of a thermo-optically induced waveguide switch fabricated in Ti diffused $LiNbO_3$. In a z-cut $LiNbO_3$ crystal the light is deflected down into the substrate under zero voltage. When a voltage is applied the undoped $LiNbO_3$ material in the gap increases in temperature and a thermo-optical waveguide is induced, coupling the light across the gap into the second Ti-diffused region.

There is an increasing variety of applications for these devices, the main ones at the present time being telecommunications and remote sensing. A useful review of the technology is given by Senior and Cusworth (1990).

8.5   Other Technologies

There are a number of complementary optical devices which have been developed as 'add on' components to the fibre and waveguide technologies. It is expected that many of these will one day be integrated into OEICs (opto-electronic integrated circuits). OEICs come in two categories:

1.   *Opto-electronic combined with electronic functions.*   These include lasers and photodetectors integrated with high-electron-mobility transistors (HEMTs), field effect transistors (FETs) and bipolar transistors for fabricating integrated laser drivers and high sensitivity receivers.
2.   *Opto-electronic combined with optical functions.*   These include lasers, photodetectors and optical amplifiers integrated with waveguide devices such as directional couplers, filters etc..

One of the most useful technologies which will become increasingly used in telephone exchanges, optical computers, optical neural network systems – in short any opto-electronic system where a large number of interconnections have to take place – is the optical holographic backplane. This technology basically allows the interconnection of outputs on one opto-electronic rack to the receivers on other racks without the use of hard wired connections thus eliminating these connectors, freeing the space between the racks for servicing purposes and allowing signal routing to be achieved by optical deflection. It is possible, for instance, to deflect or diffract a laser beam on to a particular element of an input array using either electronic or optical addressing of the hologram totally independent of the information being carried on the laser beam.

CHAPTER NINE

Liquid Crystals

9.1 Introduction

In general all matter has at least three phases, gaseous, liquid and solid. Many materials also have different phases within these classifications which are termed meso-phases. For instance, many crystals have differing structures depending on the cooling conditions under which the lattice freezes. In general the gaseous phase is totally random, only molecular order being maintained; the liquid phase has a degree of short range order and the solid phase has medium to long range order. Polycrystals and solid solutions have medium range order and single crystals have long range order. The boundary between the phases is usually characterised by a critical temperature and the material structure is more ordered below this temperature than above. The phase change is described by the conditions where the structure of the material changes and may be accompanied by large changes in physical properties. Although within a given structure many parameters (e.g. expansion coefficients, absorption, conductivity) may change more or less smoothly with temperature the material symmetry can, *a priori*, only change suddenly. The basic reason why many materials are hard to crystallize is that when the structure reaches the phase change temperature the dislocation is so great that the crystal shatters. Polycrystalline material is therefore not so much a mixture of structures but the result of stresses trying to change the structure.

While the material is in the liquid phase the structure is able to change in a non-catastrophic fashion. Although the common conception of a liquid is of an isotropic material this does not hold for a large number of materials (which are interesting for this very reason). These materials, although only having short range order, have definite short range structure and therefore have anisotropic physical and optical properties. The simplest type of spontaneously anisotropic liquids are commonly termed liquid crystals (LC). These have a series of meso-phases, (Gray 1962, Priestly 1975), and are hence also termed mesogens, and in particular have some very interesting and useful optical properties. Recent comprehensive surveys of the field have been published by Clark (1990), Chandrasekhar (1992) and Donald and Windle (1992).

9.2   Structural and Physical Behaviour

The most common liquid crystals have short rod or plate shaped structures. Depending on the temperature, these rod-like molecules exist in different meso-phases between the completely unordered, isotropic, state of a true liquid and the three-dimensional long range positional order of the solid state. At high temperatures (>100°C) the molecules are randomly oriented and the solution shows isotropic behaviour (see Figure 9.1a). As the temperature decreases the liquid crystal undergoes a phase transition into the nematic phase where the molecules take up orientational but not positional order (Figure 9.1b). When the temperature is reduced still further the liquid crystal molecules start to self-assemble into layers – the smectic phase. Up to eight smectic phases (SmA → SmH) have been identified in some liquid crystal materials, all having stable tilted structures (Gray and Leadbetter 1977), the main ones being the SmA and the SmC phases. When the long axis of the rod-shaped molecules (termed the molecular director) is oriented perpendicular to the layers the phase is called the smectic A (SmA) phase (Figure 9.1c). When the temperature drops even further the layers may contract even further and change the orientation of the rods within the layers (Figure 9.1d). The main phase in which the molecules are tilted at an angle with respect to the layer normal is termed the smectic C (SmC) phase.

The rod-like molecular structures necessary for liquid crystal operation are found in aromatic, organic materials (e.g. *p*-substituted phenylene rings linked together give common core structures). Table 9.1 lists the molecular structures and transition temperatures for some of the simpler liquid crystal materials. When the core structure is simple the nature of the end groups dominates the LC properties. With extended core materials the end groups have less influence and conversely can be tailored to give better LC properties. Common terminal groups are alkyl, -Ch3, or alkyloxy, -OCH3, and for instance if the alkyl chain is short the mesogen is usually nematic but if it is long the mesogen tends to be smectic (Gray and Leadbetter 1977). Both the meso-phase–liquid and the meso-phase–solid transition temperatures are functions of the precise molecular structure and a lot of careful work, both in terms of synthesis and measurement, has had to be undertaken to gain an

Figure 9.1   Liquid crystal alignment phases
  (a) Isotropic
  (b) Nematic
  (c) Smectic A
  (d) Smectic C

understanding of the precise relationships (see Donald and Windle 1992 for a wider view of the subject).

In the nematic phase alignment extends spontaneously over regions of the order of microns (the local axis of alignment being called the director). With no external constraints the director fluctuates in both space and time. A bulk sample can, however, be aligned by applying small electric or magnetic fields or by surface forces. In practice the bounding substrates fix the orientation of the director at each surface. The most commonly used fabrication technology uses a twisted nematic (TN) or supertwisted nematic material (STN). Twist in this instance refers to the tendency of polymers to form chains which rotate (see Figure 9.2). The degree of rotation is a function of the liquid crystal used and is locked by orienting the substrates so that the director angle at one face plate is the requisite angle from the other (90° for a TN and 270° for an STN). The switching time of a TN cell is commonly restricted to milliseconds and once the cell has been switched (by the application of a voltage) it cannot be unswitched except through its own inbuilt mechanical process.

In physical optics terms the basis of transmission through LCDs is governed by the fact that both the refractive index and the permittivity vary with the orientation of the electric vector relative to the director (Clark 1990). Thus light polarised parallel to the director experiences the extraordinary refractive index, n_e, whereas the polarisation perpendicular to the director experiences the ordinary refractive index, n_o. As a result, provided that $(n_e - n_o)d > \lambda$, where d is the thickness of the cell and λ the wavelength of the light in vacuo, the director planes guide the plane of the plane polarised light. In Figure 9.2(a) the twisted configuration guides the plane of polarisation and the device transmits when placed between crossed polarisers. When a suitable voltage is applied the director configuration distorts, leading to the case of Figure 9.2(b) where the plane of polarisation is not rotated and the crossed polarisers leave a dark field.

In materials consisting of chiral, or asymmetric, molecules the director twists about an axis perpendicular to itself giving a cholesteric meso-phase. The pitch of the twist can be of the order of the wavelength of visible light and is

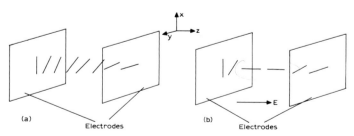

Figure 9.2   Liquid crystal device operation

Table 9.1 Molecular structure and phase data for some liquid crystal materials

Type	Molecular structure	Transition temperature (°C)		
		I → N	I or N → S	N or S → C*
Azo	CH$_3$O – Ph – N = N – Ph – C$_4$H$_9$	48	–	32
Azoxy	CH$_3$O – Ph – N = N – Ph – C$_4$H$_9$ (– O below)	77	–	44
BiPhenyl	C$_5$H$_{11}$ – Ph – Ph – CN	35	–	22.5
	C$_8$H$_{17}$ – Ph – Ph – CN	40	32.5	21
	C$_{12}$H$_{25}$ – Ph – Ph – CN	–	58	43.5
Ester	C$_5$H$_{11}$ – Ph – CO.O – Ph – OCH$_3$	42	–	29
	C$_7$H$_{15}$ – Ph – CO.O – Ph – CN	54	–	40
	C$_5$H$_{11}$ – Ph – CO.O – Ph^{Cl} – CO.O – Ph – C$_5$H$_{11}$	122	–	39
	C$_7$H$_{15}$ – Ph – CO.O – Ph – Ph – CN	224.5	–	92
TerPhenyl	C$_5$H$_{11}$ – Ph – Ph – Ph – CN	239	–	130
Stilbene	C$_2$H$_5$O – Ph – CH = CCl – Ph – C$_4$H$_9$	58	–	29
Schiff's base	CH$_3$O – Ph – CH – N – Ph – C$_4$H$_9$	47.5	–	22
	C$_4$H$_9$O – Ph – CH = N – Ph – CN	108	–	65
Tolane	CH$_3$O – Ph – C ≡ C – Ph – C$_7$H$_{15}$	54	–	39
4-di-n-heplylexyazoxybenzene (*II*)		124	95	74

Ph = benzene ring, I = Isotropic liquid, N = Nematic phase, S = smectic phase, C* = crystal phase.

very sensitive to temperature. This phenomenon allows cholesteric liquid crystals to be used as surface temperature sensors.

The smectic C phase is the most ordered and least symmetric and many of these materials possess a spontaneous permanent dipole moment (not induced by an external electric field). These materials are chiral (being unable to be superimposed on their mirror image by rotation, translation or reflection) and the lack of symmetry leads to microscopic ferroelectricity. They are termed ferroelectric liquid crystals (FLC) or SmecticC* (SmC*). Chiral structures like to settle in helical structures (each molecular layer being rotated slightly from its neighbour). The pitch of this helix differs from SmC* compound to SmC* compound and can be from 1 to 10 μm (500 to 5000 smectic layers). Since the molecules rotate about the helix with their polarisations pointing in all directions these microscopic dipole moments cancel in the bulk. If the chirality is suppressed (e.g. by squashing the material between substrates) the surface forces result in two stable states of the SmC molecular director and a bulk ferro-electric is formed. This can be used to advantage since as the two orientation states are stable (i.e. do not relax back from one to another when the voltage is removed) the FLC can be addressed using low data rate technology and/or pulsed operation. The switching times between smectic molecular layers can be of the order of 10^{-10} s (Gray and Leadbetter 1977) but the switching times for FLC devices are of the order of submicroseconds (Clark and Lagerwall 1980).

9.3   Flat-Panel Display Fabrication and Operation

A cross section schematic of the simplest liquid crystal device is shown in Figure 9.3. The cell comprises a liquid crystal layer bounded by parallel substrates which have transparent conducting electrodes and polyimide orientation layers between the substrates and the liquid crystal. The director orientation is achieved by rubbing the polyimide coated substrate with a velvet or nylon cloth is a single direction. This process imparts a directional charge to the surface which then orients the director parallel to the rubbing direction. If the rubbed plates directions are set at 90° to each other this imparts a 90°

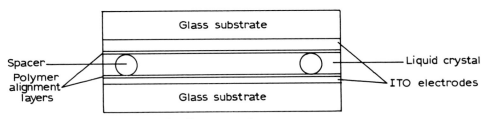

Figure 9.3   Liquid crystal cell cross section

twist to the director as it passes through the bulk of the crystal. The cell is sealed at the edges with epoxy. The spacing accuracy is crucial and is usually maintained at ¯10 μm with a tolerance of ± 0.5 μm using short lengths of optical fibre or small silica spheres. The cell is completed by the provision of polarisers and colour filters. The polarisers are oriented so that they are either crossed (transmission of the rotated polarisation with zero voltage–extinction when the voltage is applied) or perpendicular (vice-versa). Displays are lit by lighting the liquid crystal using white light and colour filters are provided to give colour. Multicolour flat panel displays are produced by fixing a matrix of blue/green/red colour filters.

When a suitable voltage is applied across the cell the liquid crystal reorients to follow the applied electric field. When the voltage is turned off the liquid crystal returns to its original state. As the polarisation of the transmitted beam rotates in phase with the crystal director the transmitted light follows the normal sinusoidal variation of extinction with angle to the polariser direction. Normal TN liquid crystals have twist angles of 90°. Supertwist nematics have twist angles of 270°, allowing higher contrast, faster response and the ability to multiplex many pixel elements in a single display. This is the mainstream technology in the early 1990s but may yet be superseded by the ferroelectric LCDs at present being investigated.

The main advantage of the ferroelectric liquid crystals is that, unlike the nematics, there are two stable states of the molecular director. This means that once an FLC is switched it stays in that orientation until the field is reversed. It also has the useful side effect that as the switching in either direction is made by applying an electric field the switching is faster than even the STN phase. In TN and STN devices the polarisation is induced by the electric field and reversing the sign of the field has little or no effect, the molecules returning to their original state through a mechanical process once the voltage has been removed. These considerations limit the response times of TN cells to the order of milliseconds while microsecond response times have been reported for FLCs (Clark and Lagerwall 1980).

9.4   Switching Technology

The main problems of producing a large area, three-colour, flat-panel liquid crystal display centre round the necessity of keeping the liquid crystal layer a constant, uniform thickness (of the order of 10 μm) and of providing three different colour dots at each pixel element. The latter requirement comes from the fact that a TV screen or computer display commonly comprises 640 rows by 480 columns, bringing the number of discrete addressable elements up to ¯10^6.

The mechanical tolerances required to keep the two substrates within tolerance are well understood and explain why 12-inch diagonal displays are of the order of 2-inch total thickness whilst the 3-inch displays can be fabricated in less than half an inch thickness.

The number of pixels required brings complications on two accounts, both of them potentially more difficult on the smaller diagonal displays than on the larger. The simpler problem is the provision of sets of complementary colour filters at each pixel element. In large area displays this can be done by fabricating a matrix of colour elements out of coloured perspex etc. This can also be done on the 0.3 mm pitch required for large area displays. However, this is an expensive, laborious and exacting process and attempts are being made to supersede this by evaporating three colour reflection filters and etching the unwanted films away using standard reprographic techniques.

The second problem is the one of addressing each pixel and this has necessitated a lot of research which has been markedly successful and which gives credit to the research teams involved.

The earlier, passive displays required the fabrication of transparent electrode stripes, the pitch being the pixel dimensions, one set being vertical and the other horizontal (see Figure 9.4). This technology makes it relatively

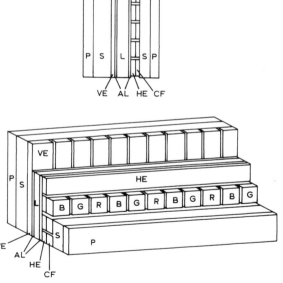

Figure 9.4   Liquid crystal switching technology: passive system

P, Thin film polarisers   S, Substrates (glass)
CF, Thin film colour filters   VE, Vertical electrodes
AL, Polymer alignment layers   HE, Horizontal electrodes
L, Liquid crystal

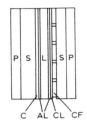

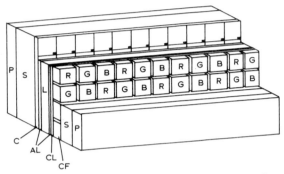

Figure 9.5   Liquid crystal switching technology: active matrix

P, Thin film polarisers   S, Substrates (glass)
CF, Thin film colour filters   CL, Conducting layer
AL, Polymer alignment layer   C, Thin film circuit
L, Liquid crystal

easy to fabricate the cell but makes it harder to operate, especially for the TN and STN cases where information rates have to be high in order to circumvent the flicker which arises if pulsed addressing techniques are used. This puts tight constraints on the drive circuitry and particularly on the thin-film deposition technology which has to have a higher repeatability than even normal IC manufacture if screens with a minimum number of unswitchable pixels are to be fabricated.

The preferred construction nowadays is called the active-matrix technique where each pixel has at least one transistor or diode fabricated at its edge (see Figure 9.5). The front panel has a continuous conducting film and the rear electrode is patterned to follow the pixel shape. The number of transistors necessary to run an active matrix makes it imperative that the possibility of failure is taken into account and that redundancy is built into the system. This has been done with a marked degree of success and it has been shown that it is just as easy to fabricate and alter the driving circuit (e.g. by laser ablation or welding through the rear transparent substrates) in this way than to provide and switch in extra circuitry on a totally separate IC chip. Readers wishing to inquire more fully into this subject are referred to the article by Clark (1990).

9.5 Applications

The properties of each of the three liquid crystal types and the phase change regions can be utilised to yield useful optical devices. It should be remembered, however, that liquid crystal cells rotate, not generate, light, and they all operate by rotating the plane of polarisation of the transmitted optical beam. When used in conjunction with other optical technologies they yield optical switches, spatial light modulators, sensors and flat panel displays.

The driving force for liquid crystal displays comes from the necessity for flat-panel TV and computer screens – principally for portable applications where the depth of a conventional CRT tube is impracticable. Other display applications are in the automotive (radio, dashboard, car-mounted TVs where both the low voltage operation and the flat panel characteristics recommend themselves, military/avionics (navigation and head-up displays) and the instrumentation industries. The FLCs are particularly advantageous for high definition TV screens (HDTV) because they can be switched far faster than the TN LCDs. The speed and bistability of FLCs also makes them favourite in data communication and information processing.

It is a reflection of the market applications that have already been realised that the sophisticated technology referred to in the last section has been funded to such a degree that its main problems have already been overcome. Active-matrix, full colour, flat-panel displays have already been manufactured and are very similar in optical/display terms to CRTs. Most of the flat-panel displays on the market in the late 1980s were ≈4-inch diagonal but larger (≈12-inch diagonal) panels have been fabricated by a number of manufacturers and although still extremely expensive are now commercially available. The main problem that remains is to make the fabrication technology consistent enough to increase the manufacturing yield to a level that will allow a profit to be made at a reasonable price per display. It is expected that the explosion in the market penetration of these devices will continue well into the 1990s.

CHAPTER TEN

Power Handling Capability of Optical Materials

10.1 Introduction

As was explained in Chapter 3, the intrinsic absorption spectrum is bounded at short wavelengths by multi-photon absorption and at longer wavelengths by band-gap absorption. Between these limits Rayleigh Scattering and Fresnel reflection define the minimum transmittance at low optical input.

It is not possible to increase the transmitted power indefinitely because of the advent of non-linear effects. Second-order effects include second harmonic generation and frequency up-conversion leading to loss of transmission of the fundamental. Second-order optical non-linear phenomena are usually absent in glasses due to inversion symmetry. Third-order non-linear effects occur at high input powers leading to spectral broadening, stimulated Raman and Brillouin scattering, non-linear switching, self-focusing and parametric four-photon mixing (see Chapter 5). To some extent the thresholds of these non-linear effects are related to the material attenuation since they are related to the intrinsic and extrinsic absorption in the material. Extrinsic processes are due to impurities or imperfections and include electronic, vibrational and surface absorption and microscopic and macroscopic scattering.

Ultimately laser induced damage (LID) to the material under observation will occur. Laser-induced damage to bulk materials is caused by a combination of dielectric and thermal breakdown which are in turn directly related to the absorption, scatter and refractive indices of the material under consideration (Wood 1975, 1986). At short pulse lengths ($<10^{-6}$ s) dielectric breakdown dominates the measured thresholds, LIDT, for low absorption materials whilst thermal effects dominate the measured values for long pulse and cw beams or when the absorption level is high (e.g. metal mirrors and detectors).

Bulk material and surface damage thresholds for a range of materials have been measured in several laboratories over a range of wavelengths and pulse lengths (Wood, 1986, 1990). Most of the published values are for the 1–100 ns and cw regimes. There is very little systematic data for the microsecond and millisecond regimes. However, both the experimentally available values and the theoretical treatments of the relationship with pulse length and spot size are consistent. These show that for transmissive materials the damage thresholds are peak power density dependent for short pulse lengths and energy density dependent for long pulses/pulse trains and cw beams. This is illustrated in Figure 10.1 where the damage threshold is plotted as a function of pulse length for air (Musal 1980), CMT detectors (Bartoli *et al.* 1975) and for fused silica as well as a more restricted selection of data points for lithium niobate. The transition point between dielectric breakdown and the onset of the thermal processes is related directly to absorption and the cooling processes (thermal conductivity, diffusivity, specific heat etc.). The damage threshold/pulse length curves for all materials therefore look like those shown in Figure 10.1, the break point being a function of the thermal properties of the particular sample material under test.

It should be reasonably easy to predict this curve for any particular material as the position of the thermally dominated portion only requires a calculation of the heat needed to melt the material under investigation and the pulsed laser damage thresholds are synonymous with electrical breakdown. However, reality rarely matches up easily and there is a wealth of data and even an annual conference to illustrate the fact. The complicating factors are as follows:

(1)   The onset of self-focussing, thus giving higher peak power and energy densities within a material than at the focal point of a lens in air.

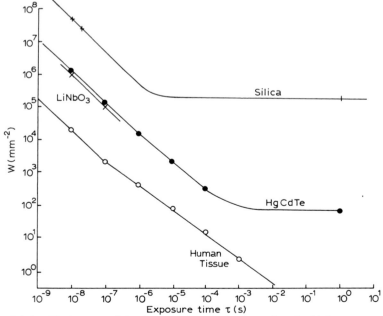

Figure 10.1   Variation of the laser induced damage threshold for a number of materials versus laser pulse length

(2) The modification of the LIDT by impurities and material imperfections.
(3) The dielectric breakdown thresholds are a function of the dielectric constant at the optical frequencies, not d.c.

These three factors have become so interwoven that although the general form of the LIDT/pulse length graph is known there is considerable variation in the experimental data measured even for fairly uniform materials.

10.2 Self-Focusing

The subject of self-focusing was discussed at length in section 3.6 and it was seen that there are a whole series of self-focusing mechanisms which operate under different time relaxation regimes. In general self-focusing is a reduction of the laser beam diameter below the value predicted from the refractive index of the unirradiated material and the laser beam/optical profile as measured in air. It occurs because the refractive index of a material is both temperature and electric field dependent:

$$n = n_{\mathrm{o}} + n_{\mathrm{T}}(T) + n_2 E^2 + \ldots$$

The temperature dependent refractive index of a transmitting material is therefore very much a function of the material impurities and the power dependent function changes as the beam constricts and yields a critical interaction length.
  For the aforesaid reasons it becomes difficult to reconcile the different experimentally measured data as even such factors as the depth of the focus from the surface of a damage sample can affect the result. However, although it has been seriously doubted whether any damage measurement experiment has ever been unaffected by self-focusing it is only a focusing effect and the cause of laser induced damage is still thermally induced melting, cracking or dielectric breakdown.

10.3 Laser Induced Damage

10.3.1 Thermal Damage

The simplest case of thermally induced damage occurs at the surface of absorbing materials. In this context the surface absorbs energy from the laser beam and, although the heat is partially dissipated by conduction and radiation, if the flux density is great enough it heats the surface (Prokhorov 1990). When the temperature rise at the centre of the beam is sufficiently high the surface melts. A simple calculation of the parameters involved leads to the following equation (Gibbs and Wood 1976):

$$E_d f(\tau) = (T_M - T_A)(\pi k C\rho)^{\frac{1}{2}}/A$$

where E_d = laser beam peak energy density
$\quad C$ = specific heat
$\quad \rho$ = density
$\quad k$ = thermal conductivity
$\quad A$ = absorption
$\quad T_M$ = melting point
$\quad T_A$ = ambient temperature
$\quad f(\tau)$ = laser pulse shape factor

A simple thermal analysis of the situation leads to the conclusion that the time dependent function for most simple pulse shapes is $\sim\tau^{\frac{1}{2}}$. This agrees with experimentally measured data for both long and short pulses (Wood 1986, Figuera *et al.* 1982, Marrs *et al.* 1982).

A complication arises in the calculation of the exact values for the LIDT as most of the relevant parameters (conductivity, density and specific heat) are also a function of temperature (Arnold 1984).

In the special case of metal surfaces, where all the absorption is in an extremely thin skin depth, the above equation becomes easy to illustrate. Table 10.1 lists the relevant parameters as well as the calculated and highest measured (10.6 µm) LIDT values. These values have been normalised to 50 ns from data gained at a series of pulse lengths (Wood 1993). It will be seen both from this table and from Figure 10.2 that there is very good agreement between the values, notwithstanding the uncertainty of the parametric variation with temperature. The only really variable factor in the equation is the absorption because, as we shall see in section 10.4.1, this can be markedly affected by the surface finish. It is, however, possible to calculate the minimum value for this absorption both from measurements of the complex refractive index and theoretically using Drude Theory.

In the case of transmitting material (NB: no material has zero absorption) laser-induced melting rarely takes place. Instead as the heat is absorbed in a cylinder of material and flows outwards from this cylinder a cylindrical strain build up and the material eventually cracks, also cylindrically unless the cooling is asymmetric.

In the case of partial transmitters, such as gallium arsenide or silicon at 10.6 µm, the precise damage mechanism is affected by the presence of impurities or dislocations. If a crystal has a large dislocation density and is situated in a fairly high power laser beam (but below the thermal damage level) the dislocations will gradually be attracted by the electromagnetic field

Table 10.1 Laser-induced damage thresholds for metal mirrors

Mirror	Melting point $(T_m \,°C)$	Thermal conductivity (K)	Density ρ $(gm\ cm{-3})$	Specific heat, C $(cal\ gm^{-1})$	Absorption $(A\%)$	LIDT J cm^{-2} Rel. to Cu	Calc	Highest measured
s-cCu	1083	0.92	8.92	0.092	0.3[a]	2.15	130	135
Dt Cu	1083	0.92	8.92	0.092	0.4	2.0	120	90
Dt Cu	1083	0.92	8.92	0.092	0.8[b]	1.0[b]	60	60
Dt Cu/Au	1062[d]	0.92[c]	8.92[c]	0.092	1.0	0.78	47	40
Dt Au	1062	0.70	17.0	0.031	1.0	0.54	32.5	37.5
Au/Ni/Cu	1062[d]	0.13[f]	8.9[f]	0.106[f]	1.0	0.30	18	15
Al	660	0.50	2.7	0.215	2.0	0.08	5.0	3.3
Ni	1455	0.13	8.9	0.106	2.0	0.02	1.2	–
Ag	961	1.00	10.5	0.057	1.0	0.64	38.5	40
Cr	1890	0.21	7.2	0.107	0.5	1.2	70	–
Mo	2620	0.35	10.2	0.060	3.0	0.34	20.5	23
Mo eb	2620	0.35	10.2	0.060	2.0[a]	0.5[a]	30	30

[a] estimated from measured LIDT, [b] standard, [d] Au, [e] Cu, values [f] Ni, 50ns CO$_2$TEA

s-c, single crystal; Dt, diamond turned; eb, electron beam finished

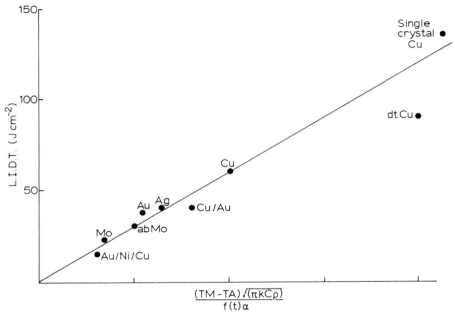

Figure 10.2   Calculated and measured LIDTs for metal mirrors

towards the beam and in time will join together forming larger discontinuities. When these discontinuities are massive enough they disturb the
heat flow and the material under investigation will suddenly damage due to
melting. This has been shown to exist in the form of a narrowing cylinder
from front to back of such a crystal.

10.3.2   Dielectric Breakdown

Laser-induced damage occurs in bulk materials at electromagnetic field
strengths which induce dielectric breakdown of the material. Massive
insulators have a.c. dielectric breakdown strengths, V_B, of about 1 to 5
MV cm^{-1} (Maissel and Glang 1970, Fradin and Bass 1973). These electromagnetic field strengths can be induced by high-power density radiation at
optical power densities, P_D, of the order of 50–1000 MW mm^{-2}. The
relationship between the a.c. dielectric breakdown strength and the laser-
induced damage threshold, LIDT, is:

$$P_D = V_B^2/Z_1 = V_B^2 n/Z_0$$

where Z_1 and Z_0 are the impedance of the dielectric and free space respectively,
n is the refractive index of the material and

$$P_D = E_p/\tau A$$

where E_p is the total energy in the pulse, A is a measure of the beam cross sectional area and τ is the laser beam temporal parameter.

Both A and τ need careful definition and measurement as very few focused laser spots are square. In the case of a perfectly focused Gaussian beam the area parameter becomes πb^2 where b is the $1/e^2$ intensity diameter. The equivalent beam pulsewidth is the full width of a temporally square pulse, the full width at half maximum (FWHM) of a perfectly triangular pulse or some aggregation of the two in the case of most laser pulses. In practice most Q-switched pulses are asymmetric, having a faster rise time than fall time but even this is not sacrosanct. It is worth noting at this point that the relationship between the breakdown strength and the optical density is independent of wavelength except through the refractive index, although is should be realised that the focused spot size will vary with wavelength ($b^\sim\lambda^2$).

A list of representative values are shown in Table 10.2. This table includes measured dielectric breakdown fields, refractive indices and both calculated and measured laser-induced damage thresholds. The true bulk LIDTs, corresponding to dielectric breakdown can only be measured inside optically perfect materials. Any inhomogeneity in the bulk material tends to produce local higher power and energy densities. As these inhomogeneities include voids, absorptive inclusions, surface cracks, grooves and pores it is no wonder that the commonly quoted ranges for the measured LIDT are so wide.

10.4   Surfaces and Coatings

10.4.1   Surface Effects

The bulk dielectric breakdown value is rarely measured in the windows, mirrors or components comprising normal laser systems as damage to one or other of the component surfaces usually occurs at lower laser flux densities. Measured damage thresholds are also apparently higher for the front than for the rear surfaces of thin transmitting components. In fact, the electromagnetic breakdown intensities are equal for both surfaces once reflections at the surface and the phase shifts of the optical field are taken into account. For a good sample with refractive index, n, the relationship between the exit and entrance window LIDTs is $4n^2/(n + 1)^2$ (Stickley 1973). In addition, a standing wave is formed which has a maximum constructive power density at a distance $\lambda/2$ (where λ is the laser wavelength) in front of the surface. As this point in inside the rear surface of an optical component and in front of the front surface the plasma initiated at the breakdown damages the material in the former case while giving a modicum of protection in the latter.

Table 10.2 Comparisons of measured and calculated laser-induced damage thresholds, refractive indices and dielectric breakdown fields

Material	Wavelength (μm)	Dielectric breakdown threshold (MVcm^{-2})	Calculated LIDT (MW mm^{-2})	Measured LIDT (MW mm^{-2})	Refractive index n
Sodium chloride	DC	1.50	86	–	–
	0.69	2.3	202	–	1.55
	1.064	2.3	202	300	1.54
	10.6	1.95	145	43	1.50
Potassium chloride	DC	1.00	37	–	–
	1.064	0.86	28	70	1.49
	10.6	1.07	43	50	1.45
Silicon	2.94	0.13	1.5	–	–
	10.6	0.75	48	12	3.4
Gallium arsenide	2.94	0.1	0.8	–	3.32
	10.6	1.4	165	8	3.27
Quartz	1.064	5.0	960	>200	1.45
BSC glass	1.064	4.7	880	>1000	1.50

As has already been noted, the true bulk LIDT corresponding to dielectric breakdown is rarely reached because of the influence of discontinuities. This is particularly true at the surface where scratches, digs and pits predominate. These discontinuities act as both dielectric field enhancement sites and as absorption traps, both for intrinsic and for extrinsic impurities. The dielectric enhancement treatment has been well treated by Bloembergen (1973) and this is illustrated in Figure 10.3. It will be realised from this that the highest surface damage thresholds are obtained only with near perfect surfaces. This puts extremely high tolerances on the surface finish of optical components for use with high power lasers. This is exemplified by the example shown in Figure 10.4 where the variation of the surface damage threshold is shown as a function of the surface finish for a specimen of homogeneous glass (Wood 1975). It will be seen from this example that the magnitude of the surface scratches etc. lowers the LIDT consistently (the precise shape of the curve is a log/log plot). When the surface contains no structure greater than about $\frac{1}{10}$ of

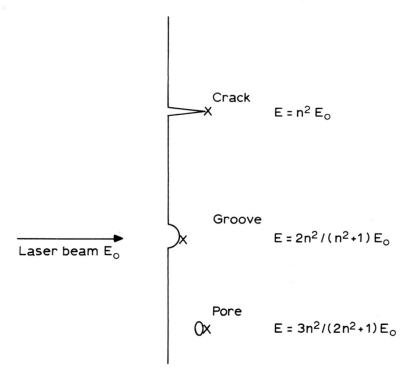

Figure 10.3   Enhancement of the electric field, V, by cracks, grooves and pores (Bloembergen 1973)

Table 10.3   Summary of laser induced damage thresholds

Material	Front surface (MWmm^{-2})	Inclusions (MWmm^{-2})	Bulk-void (MWmm^{-2})	Intrinsic (MWmm^{-2})	Rear surface (MWmm^{-2})	RI (n)
Nd : YAG	8 → 53	–	10	34 → 82	11 → 45	1.83
Nd : YALO	11 → 26	–	10	–	11 → 86	1.8
Nd : CaWO$_4$	10 → 30	3 → 4	10	10 → 50	10 → 27	1.7
Nd : glass	10 → 350	1 → 10	10 → 18	12 → 350	12 → 180	1.5
KD*P	7 → >43	1 → 5	15	10 → 180	6 → 170	1.5
KDP	10 → >240	1 → 3	10	10 → 570	10 → 185	1.5
ADP	4 → >240	1 → 3	10	5 → 220	14 → >45	1.5
LiNbO$_3$	1 → 111	1 → 5	5	1 → 53	5 → 10	2.2
Calcite	5 → 30	1 → 5	5	5 → 20	5 → 15	1.65
Quartz	12 → 300	–	10	12 → 310	20 → 52	1.45
BSC glass	5 → 46	5 → 10	10	10 → 5000	10 → >180	1.5
Flint glass	9 → 20	3 → 6	6	6 → >60	10 → 20	1.7
Fused silica	13 → 240	5 → 10	10	20 → 5000	20 → >180	1.45

Measurements made at 1.064 µm
Measurements normalised to τ = 10 ns

Table 10.4 Summary of laser-induced damage
thresholds for infra-red materials

Material	L. I. $(MWmm^{-2})$ minimum	D. T. $(MWmm^{-2})$ maximum	RI (n)
Germanium	0.6	11	4.0
Zinc sulphide	2	9	2.15
Zinc selenide	0.8	15	2.45
Gallium arsenide	1.1	8	3.27
Cadmium telluride	1.1	2.9	2.65
Silicon	0.4	13	3.42
Sodium chloride	30	43	1.50
Potassium chloride	10	50	1.45
Diamond	22	>40	2.42
As_2S_3	1	2.2	2.38

Measurements made at 10.6 μm
Measurements normalised to $\tau = 100$ ns

the wavelength of the laser radiation the surface threshold is found to be of the same order as the bulk LIDT. This has also been observed in recent LIDT measurements at a number of wavelengths on polycrystalline CVD grown diamond (Klein 1992, Sussmann 1993).

Tables 10.3 and 10.4 show a summary of laser-induced damage thresholds of a range of optical materials at 10.6 μm and 1.064 μm. It is instructive to note that whilst measurements made at 1.064 μm can be used to differentiate front surface, rear surface, inclusions, voids and intrinsic damage, those at 10.6 μm can only be used to differentiate between thermal damage (usually low values of $1-5$ J cm^{-2}) and dielectric breakdown ($10-50$ MW mm^{-2}).

10.4.2 Coatings

Most optical systems necessitate the surfaces of the components to be coated. This may be to prevent stray reflections interfering with the quality of the beam, to achieve maximum transmission, to beam split a given amount of the beam to a feed-back system or sensing loop or to achieve maximum reflectance in a mirror system. All these coatings are, of necessity, placed on the surfaces of the components. They are therefore liable to be affected by the quality of the surface finish and in practice these coatings constitute the actual limitations of the system. Most high power laser systems are therefore run at beam powers such that the maximum laser power density at any point of the optical system is lower than the coating LIDT. In practice a margin of safety has also to be introduced as coatings often deteriorate during life and this has to be taken into account if a long life system is required. In addition to

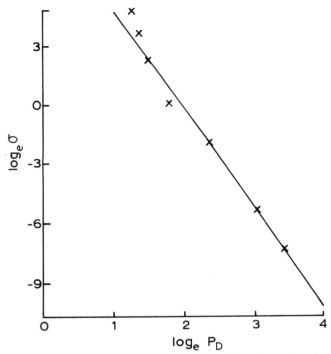

Figure 10.4   Variation of surface damage threshold with surface finish

ensuring that the component surface is flaw free before coating it is also sensible to use coating designs such that the standing wave voltage ratio (the sinusoidal wave train) is such that it is not a maximum either at the surface of the coating system or at the point of change of the coating material. This subject is a very large one and the reader is directed towards excellent treatments by Heavens (1965), Maissel and Glang (1970) and MacLeod (1986). A lot of research has taken place over the years into coating techniques, designs and materials. It is nevertheless sufficient to say that the requirements on optical coatings grow ever tighter and the search for higher and higher quality will go on until not only will the coating quality be perfect (no scratches, digs or pits) but it will also protect the surfaces from extraneous influences and will allow the optical components to be placed within a high power laser resonator without fear of laser-induced damage. For these reasons the coating materials have progressed from the early days of the fluorides and sulphides through the tougher oxides and into the field of carbide (Brierley 1991) and nitride (Wood 1990) coatings. The nitride coatings in particular offer minimal absorption, scratch-resistant coatings which can be deposited to yield high efficiency/high damage threshold coatings over wide spectral ranges.

10.5   Scaling Laws

It is necessary when designing an optical system to first specify the required optical characteristics of the system, then to design it with an eye to the characteristics of the available optical materials and lastly to make sure that the optical power and energy densities to which it is to be subjected are within the tolerances of the laser damage thresholds of the materials specified and that non-linear effects do not bring unwanted changes into the final system. When determining whether or not the system can withstand the high powers of radiation it is useful to have data gained at similar wavelengths on similar materials and at similar pulse lengths. This is available in some cases but as the number of wavelengths used increases it is frequently necessary to extrapolate between similar data gained at another wavelength/pulse length or spot size.

Unless the sample absorption is extremely low at the wavelengths of interest the damage threshold will be directly related to the threshold. Only dielectric breakdown of all the laser-induced damage mechanisms is totally independent of the wavelength, as long as all other conditions are kept constant and/or the focal spot is large. In the general case E_D decrease with increasing λ.

As the breakdown voltage varies with an approximate relationship, $V_B \sim \tau^{-0.25}$ the LIDT can be extrapolated fairly well at short pulse lengths. This has been proven for a range of materials (Bettis *et al.* 1976). Variations from this relationship come about because of the presence of impurities (a $\tau^{-0.5}$ dependence has been widely shown to apply to surface damage at short pulse lengths). A $\tau^{-0.3}$ dependence applies to coatings on good well-polished substrates (Guenther 1984).

The distribution and the nature of surface defects leads to a variation of the laser-induced damage threshold across a surface. In particular this leads to an apparent spot size dependence of the LIDT for optical coatings. This is because coating defects tend to be circular and the focused spot shape of a laser beam tends to be Gaussian. As a result of this there is a probability factor where the defect spacing is greater than the spot size because there is a well defined probability of the laser spot overlapping the coating defect:

$$P = 1 - e^{(-\lambda A)}$$

where λ is the wavelength of the radiation and A the area of the defect.

In practice this becomes a useful measure of both the LIDT and the coating defect density. The phenomenon can be illustrated by reference to Figure 10.5 which shows plots of the percentage of the area damaged versus the damage threshold as a function of both spot size and laser pulse repetition

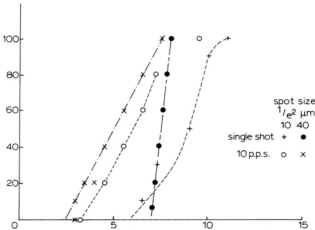

Figure 10.5   Percentage of areas damaging versus irradiation level for different laser spot sizes and pulse repetition frequencies

frequency. It will be observed that the graphs are steeper the larger the spot size. It will also be observed that in this particular case the measurements made at 10 pps gave lower LIDTs than those made at 1 pps. This illustrates that there was an appreciable absorption in the coating under investigation and that at 10 pps the repetition rate there was not enough time for the cooling mechanisms to keep the substrate at ambient temperature.

In the case of extremely small spot testing it has been observed that it is often the total pulse energy rather than the energy density which gives rise to damage. This occurs when the heat has time to diffuse from the centre to the edge of the spot during the pulsewidth. The time needed to diffuse the heat out of a diameter d is given by

$$t_d = d^2/4D$$

where D is the material diffusivity coefficient.

The heat diffuses out of the spot if $\tau > t_d$, i.e. $t_d/d^2 < \frac{1}{4}D$

When this inequality is reached the laser damage threshold becomes the total energy in the pulse spread over the entire sample area, rather than the laser spot energy density and therefore, for long pulse lengths, the LIDT becomes pulse length independent (i.e. damage occurs at constant power).

References

Agrawal, G. P. and Boyd, R. (1992) *Contemporary Non-Linear Optics.* Academic Press.

Al-Saidi, A. and Harrison, R. G. (1985) 'Wavelength dependence of high-efficiency second harmonic generation in $CdGeAs_2$', *Appl. Phys. B* **36**, 17.

Arnold, G. S. (1984) 'Absorptivity of several metals at 10.6 μm: Empirical expressions for the temperature dependence computed from Drude theory,' *App. Op.* **23**(9), 1434.

Bartoli, F., Esterowitz, L., Kruer, M. and Allen, R. (1975) 'Thermal modelling of laser damage in 8–14 μm HgCdTe photoconductive and PbSnTe photovoltaic detectors,' *J. Appl. Phys* **46**(10), 4519.

Bettis, J. R., House, R. A. and Guenther, A. H. (1976) *Spot Size and Pulse Duration Dependence of Laser-Induced Damage*, NBS Special Publication, Vol. 462, 338.

Bliss, E. S. (1971) 'Pulse duration dependence of laser damage mechanism,' *Optoelectronics* **3**, 99.

Bloembergen, N. (1973) 'Role of cracks, pores, and absorbing inclusions on laser induced damage threshold at surfaces of transparent dielectrics,' *Appl. Op.* **12**(4), 661.

Boechat, A. A. P., Su, D., Hall, D. R. and Jones, J. D. C. (1991) 'Bend loss in large core multimode optical fibre beam delivery systems,' *Appl. Op.* **30**(3), 321.

Born, M. and Wolf, E. (1975) *Principles of Optics.* Pergamon.

Brierley, C. J. (1991) 'The growth, properties and applications of CVD diamond,' *GEC Rev.* 7(2), 87.

Bubenzer, A., Dischler, B. and Nyaiesh, A. (1981) *Optical Properties of Hydrogenated Amorphous Carbon (-C:H) – a hard coating for ir optical elements*, NBS Spec. Pub. 638, 477.

Butcher, P. and Cotter, D. (1990) *The Elements of Non-Linear Optics.* Cambridge University Press.

Clark, M. G. (1990) *Liquid Crystal Devices. Encyclopaedia of Physical Science and Technology: Yearbook.* Academic Press.

Clark, N. A. and Lagerwall, S. T. (1980) 'Submicrosecond bistable electro-optical switching in liquid crystals,' *Appl. Phys. Lett.* **36**, 899.

Chandrasekhar, S. (1992) *Liquid Crystals.* Cambridge University Press, 2nd edn.

Cullis, A. G. and Canham, L. T. (1991) 'Visible light emission due to quantum size effects in highly porous crystalline Si,' *Nature* **353**, 335.

Davit, J. (1970) *Filamentary Damage in Glasses.* NBS Spec. Pub. 341, 37.

Dekker, A. J. (1960) *Solid State Physics.* Macmillan.

DeMicheli, M. and Ostrowsky, D. (1990) 'Non-linear integrated optics,' *Physics World* **3**, 56.

Dereniak, E. L. and Crowe, D. G. (1984) *Optical Radiation Detectors.* Wiley, New York.

Ditchburn, R. W. (1958) *Light.* Blackie & Son.

Donald, A. M. and Windle, A. H. (1992) *Liquid Crystalline Polymers.* Cambridge University Press.

Donaldson, R. W. and Edwards, J. G. (1983) *Concise Databook of Optical Materials.* NPL.

Eckart, R. C., Masuda, H., Fan, Y. X. and Byer, R. L. (1990) 'Absolute and relative nonlinear optical coefficients, measured by phase-matched second harmonic generation,' *IEEE J. Quant. Elec,* **26**(5), 922.

Edwards, J. G. (1967) 'An accurate carbon-cone calorimeter for pulsed lasers,' *J. Phys. E. Sci. Inst.* **44**, 835.

Edwards, J. G. and Jefferies, R. (1964) 'Response time of F4000 BiPlanar photocell in various holders,' *J. Phys. E. Sci. Inst.* **2**(2), 126.

Edwards, J. G. (1970) 'A glass disc calorimeter for pulsed lasers,' *J. Phys. E. Sci. Inst.* **3**, 454.

Edwards, J. G., Roddie, A. G. and Smith, P. A. (1983) 'Improved electrodes for photon drag detectors,' *J. Phys. E. Sci. Inst.* **16**, 526.

Figuera, J. F., Thomas, S. J. and Harrison, R. F. (1982) *Damage Thresholds to Metal Mirrors by Short-pulse CO_2 Laser Radiation*, NBS Special Pub. 638, 229.

Fradin, D. W. and Bass, M. (1973) *Studies of intrinsic optical breakdown*, NBS Special Publication, 387, 225; *Appl. Phys. Lett.* **22**, 157.

Fradin, D. W. and Bua, D. P. (1974) 'Laser induced damage in ZnSe,' *Appl. Phys. Lett.* **24**(11), 555.

Friend, R., Bradley, D. and Holmes, A. (1992) 'Polymer LEDs,' *Phys. World*, Nov., 42.

Garlick, G. F. J. and Gibson, A. F. (1948) 'Luminescent materials,' *Proc. Phys. Soc. (London)*, **A601**, 574.

Garlick, G. F. J. (1949) *Luminescent materials.* Oxford University Press.

Ghatak, A. K. and Thyagarajan, K. (1989) *Optical Electronics.* Cambridge University Press.

Gibbs, R. and Wood, R. M. (1976) *Laser induced damage of mirror and window materials at 10.6 μm*, NBS Special Pub. 462, 181.

Gibbs, R. and Lewis, K. L. (1978) 'Laser calorimeter for the determination of the absorption coefficients of thin 10.6 μm windows,' *J. Phys. E. Sci. Inst.* **11**, 304.

Goodwin, D. W. and Heavens, O. S. (1968) 'Doped crystal and gas lasers,' *Prog. Phys.*, **XXXI**(2), 777.

Gorton, E. K. (1986) *Linear and Nonlinear Properties of $HgIn_2Te_4$*, 7th National Quantum Electronics Conference, 1985. HMSO, 1215.

Gray, G. W. (1962) *Molecular Structure and the Properties of Liquid Crystals.* Academic Press.

Gray, G. W. and Leadbetter, A. J. (1977) 'Liquid crystals: what makes a mesophase?' IOP, *Physics Bulletin*, 28.

Greenham, A. C. G., Nichols, B. A., Wood, R. M., Nourshargh, N. and Lewis, K. L. (1993) 'Design and realization of optical interference filters with continuous refractive index modulation,' *Optical Engineering*, to be published.

Guenther, K. H. (1984) '1.06-μm laser damage of thin film optical coatings: a round-robin experiment involving various pulse lengths and beam diameters,' *Appl. Op.* **23**(21), 3743.

Handbook of chemistry and physics (1989) Tables of physical and chemical properties. Chemical Rubber Company Press Inc., Florida.

Harris, D. C. (1992) *Infra-red window and dome materials.* SPIE, TT10.

Heavens, O. S. (1965) *Optical Properties of Thin Solid Films*, Dover Press.

Hecht, J. (1984) 'III–V semi-conductor laser diode review,' *Lasers and Applications*, **1**, 61.

Hecht, J. (1991) 'World's smallest lasers,' *Lasers and Optronics*, **12**, 6.

Higgins, T. V. (1992) 'Non-linear crystals: where the colours of the rainbow begin,' *Laser Focus World*, **1**, 125.

Jenkins, F. A. and White, H. E. (1957) *Fundamentals of Optics.* McGraw-Hill.

Klein, C. A. (1992) *Laser induced damage thresholds of diamond*, SPIE 1624.

Kolinsky, P. V. and Jones, R. J. (1989) 'Advances in materials for non-linear optics,' *GEC J. Research*, **7**(1), 46.

Kressel, H. (1982) *Semi-Conductor Devices for Optical Communications*, 2nd edn. Springer-Verlag, Berlin.

Kushida, T., Marcos, H. M. and Geusic, J. E. (1968) 'Laser transition cross-section and fluorescent branching ration for Nd^{3+} in YAG,' *Phys. Rev.* **167**(2), 289.

Land, E. H. (1951) 'Manufacture of Polaroid,' *J. Opt. Soc. Am.*, **41**, 957.

Lee, T. P. (1989) 'Diode-laser developers look to broadband optical communications,' *Laser Focus World*, **8**, 129.

Lin, J. T. and Chen, C. (1987) 'Choosing a non-linear material,' *Lasers and Optronics*, **11**, 59.

Maissel, L. I. and Glang, R. (1970)   *Handbook of Thin Films*. McGraw-Hill.

Maiman, T. H. (1960)   'Stimulated optical radiation in ruby,' *Nature*, **187**(8), 493.

MacLeod, H. A. (1986)   *Thin-Film Optical Filters*. Adam Hilger.

Marrs, C. D., Faith, W. N., Dancy, J. H. and Porteus, J. O. (1982)   'Pulsed laser induced damage of metals at 492 nm,' *App. Opt.* **21**(22), 4063.

Messenger, H. W. (1992)   Detector handbook, *Laser Focus World*, March, p. 107.

Midwinter, J. and Guo, Y. (1992)   *Optoelectronics and Lightwave Technology*. Wiley.

Meredith, G. R. (1986)   'Appraisal of nonlinear organic optical materials,' *SPIE* **704**, 234.

Musal, H. M. (1980)   *Pulsed laser initiation of surface plasma on metal mirrors*, NBS Special Publication 620, 227

Moses, A. J. (1971)   *Handbook of Electronic Materials Vol. 1: Optical Materials Properties*. Plenum Press.

Najafi, S. I. (1992)   *Introduction to Glass Integrated Optics*, Artech House, Boston.

Nayar, B. K. and Winters, C. S. (1990)   'Organic second-order non-linear optical materials,' *Opt. & Quantum Elec.* **22**, 297.

Nourshargh, N., Starr, E. M. and McCormack, J. S. (1985)   'Plasma deposition of integrated optical waveguides,' *Proc. SPIE* **578** 95.

Nourshargh, N., Starr, E. M. and McCormack, J. S. (1986)   'Plasma deposition of GeO_2/SiO_2 and Si_3N_4 waveguides for integrated optics,' *Proc. IEE* **133**(J4), 264.

Olver, A. D. (1992)   *Microwave and optical transmission*, Wiley, Chichester.

Parker, D. (1990)   'Optical detectors: research to reality,' *Physics World* **3**, 52.

Phillips, F. C. (1962)   *An Introduction to Crystallography*. Longmans.

Pini, R., Salembeni, R., Matera, M. and Chinlon, Lin (1983)   'Wideband frequency conversion in the UV by nine orders of stimulated Raman scattering in a XeCl laser pumped multimode silica fibre,' *Appl. Phys. Lett.* **43**(6), 517.

Priestly, E. B. (1975)   *Introduction to Liquid Crystals*. Plenum Press.

Prokhorov, A. M., Konov, V. I., Ursu, I. and Mihailescu, I. N. (1990)   *Laser Heating in Metals*. Adam Hilger.

Randall, J. T. (1939)   'Luminescence,' *Trans. Faraday Soc.*, **35, 2**; *Proc. Roy. Soc. (London)*, **A170**, 272.

Rawson, A. (1991)   *Glasses and their Applications*, Institute of Materials.

Rundle, W. J. and Stitch, M. L. (1969)   'The ruby laser: its present and future,' *Laser Focus* (**3**), 27.

Savage, J. A. (1984) *The infra-red Applications of Chalcogenide Glasses*, HMSO.

Savage, J. A. (1985) *Infra-red Optical Materials and their Anti-Reflection Coatings*. Adam Hilger.

Schawlow, A. L. and Townes, C. H. (1958) 'Infra-red and optical lasers,' *Phys. Rev.* **112**(12), 1940.

Senior, J. M. and Cusworth, S. D. (1990) 'Wavelength division multi-plexing in optical fibre sensor systems and networks: a review,' *Optics and Laser Technology*, **22**(2), 113.

Shand, M. L. and Walling, J. C. (1982) 'Excited-state absorption in the lasing wavelength region of alexandrite,' *IEEE J.Q.El.* **QE18**(7), 1152.

Smakula, A. (1962) 'Synthetic crystals and polarising materials,' *Optica Acta* **9**, 205.

Soileau, M. J., Franck, J. B. and Veatch, T. C. (1980) *On Self-focussing and Spot Size Dependence of Laser-induced Breakdown*, NBS Spec. Pub., **620**, 385.

Soileau, M. J., Williams, W. E., Mansour, N. and Van Stryland, E. W. (1989) 'Laser induced damage and the role of self-focussing,' *Opt. Eng.* **28**(10), 1133.

Stein, M. L., Aisenberg, S. and Bendow, B. (1981) *Studies of Diamond-like Carbon Coatings for Protection of Optical Components*, NBS Special Pub. **638**, 482.

Stickley, C. M. (1973) *The ARPA Program on Optical Surface and Coating Science*, NBS Special Publication, **387**, 3.

Stolen, R. J. (1979) 'Non-linear properties of optical fibres,' *Optical Fibre Communications*, Academic Press.

Sussmann, R. (1993a) 'Properties and applications of bulk CVD polycrystal-line diamond,' *Proceedings of the 4th European Conference on Diamond*, Portugal, Sept., Elsevier.

Sussmann, R. (1993b). Private communication.

Sussmann, R., Wort, C. J. H., Scarsbrook, G. A., Sweeney, C. and Wood, R. M. (1993) 'Laser damage testing of CVD grown diamond windows,' *Proceedings of the 4th European Conference on Diamond*, Portugal, Sept., and SPIE: Proceedings of the 25th Boulder Damage Conference, Boulder, Oct., Elsevier.

Sweeney, C. G., Cooper, M. A., Wort, C. J. H., Scarsbrook, G. A. and Sussmann, R. (1993) *Proceedings of the 4th European Conference on Diamond*, Portugal, Sept., Elsevier.

Taylor, J. R. (1992) *Optical Solitons – Theory and Experiment*, Cambridge University Press.

Waxler, R. M., Cleck, G. W., Malitson, I. H., Dodge, M. J. and Hahn, T. A.

(1971) 'Optical and mechanical properties of some neodymium-doped laser glasses,' *J. Research NBS* **75A**(3), 163.

Wolfe, W. L., Ballard, S. S. and McCarthy, K. A. (1985)   *Refractive Index of Special Crystals and Certain Glasses.* NBS.

Wood, R. M., Taylor, R. T. and Rouse, R. L. (1975) 'Laser damage in optical materials at 1.06 µm,' *Optics and laser technology* **6**, 105.

Wood, R. M., Sharma, S. K. and Waite, P. (1984)   *Optical Characteristics of Germanium at 10 microns,* NBS Special Publication **669**, 33.

Wood, R. M. (1986)   *Laser Damage in Optical Materials,* Adam Hilger.

Wood, R. M. (1990)   *Selected Papers on Laser Damage in Optical Materials,* SPIE Milestone Series, MS24.

Wood, R. M., Nichols, B. A., Greenham, A. C., Nourshargh, N. and Lewis, K. L. (1990) 'Fabrication and characterisation of microwave plasma assisted chemical vapour deposited dielectric coatings,' *SPIE* **1441**, 316.

Wood, R. M., (1993)   'Surface finish of metal mirrors and the laser damage threshold,' *SPIE Conference on Metal Mirrors,* London, 1992. SPIE.

Wort, C. J., Sweeney, C. G., Scarsbrook, G. A., Sussman, R. and Whitehead, A. (1993) 'Optical properties of bulk polycrystalline CVD diamond,' *Proceedings of the 4th European Conference on Diamond,* Portugal, Sept., Elsevier.

Yariv, A. (1984)   *Optical Waves in Crystals.* John Wiley.

Yazaki, T., Kanatani, K. and Sakamoto, S. (1971) 'Longitudinal electro-optic effect in oblique-cut strontium barium niobate plates,' *Japan J. Appl. Phys.,* **10**, 522.

Young, C. G. (1969)   Glass lasers, *Proc. IEEE* **57**(7), 1267.

Zverev, G. M., Levchuk, E. A., Maldutis, E. K. and Pashov, V. A. (1969) 'Auto-focussing of laser radiation in active materials and non-linear crystals,' *Soviet Physics Solid States,* **11**(4), 865.

Zverev, G. M. and Pashkov, V. A. (1970)   'Self-focussing of laser radiation in solid dielectrics,' *Soviet Physics JETP* **30**(4), 616.

Index

The Unique
ONE-RECIPE
THREE-MEAL
Family Cookbook

The Unique
ONE-RECIPE
THREE-MEAL
Family Cookbook

SARA LEWIS

Sebastian Kelly

This edition published in 1997 by
Sebastian Kelly
2 Rectory Road
Oxford OX4 1BW

© Anness Publishing Limited 1996

Produced by Anness Publishing Limited

ISBN 1 901688 34 8

A CIP catalogue record for this book is available from the British Library

Publisher: Joanna Lorenz
Project Editors: Judith Simons and Emma Wish
Designer: Sue Storey
Special Photography: John Freeman
Stylist: Judy Williams
Home Economist: Sara Lewis

This book was previously published as part of a larger
compendium, *Cooking for Babies and Toddlers*

Printed in Singapore by Star Standard Industries Pte. Ltd.

1 3 5 7 9 10 8 6 4 2

MEASUREMENTS
Measurements have been provided in the following order:
Metric, Imperial and American. It is essential that methods
of measurement are not mixed within a recipe. Where a
conversion results in an awkward number, this has been rounded
for convenience, but will still provide a successful result.

CONTENTS

INTRODUCTION

Eating together as a family takes on a new dimension as your family grows and there's a baby, fussy toddler and Mum and Dad to feed together. Rather than attempting to cook three different meals or to cook one very simple meal you know the children will like, opt instead to cook three meals

from one basic set of ingredients; a simple baby dinner, eye-catching meal for even the fussiest of toddlers and a spicy dinner for the grown-ups.

The recipes are aimed at a baby aged nine months and over, a toddler or small child aged eighteen months to four years and two adults.

Family Meals

Family meal times should be a pleasure but they can all too often turn into a battleground. Whatever family rules you have about table manners it's important that all the family knows about them and that you are consistent. What matters to some families may not be important to you. If you want your children to stay at the table until everyone has finished, make sure all the family understands this. Some parents may feel it's more relaxing if the children leave the table once they have finished, leaving the adults to eat the rest of their meal in relative peace. Whatever you decide, stick to it, don't be brow beaten by well-meaning grandparents or friends.

Try to eat together as a family at

Left: *Part of the fun of eating together is sharing the preparation, too. Children love helping in the kitchen, and this involves them in the meal from the very start.*

Below: *A relaxed but polite atmosphere will enhance enjoyment and consequently encourage good eating habits.*

Similarly, beware of tiny hands reaching out from the highchair to grab a mug of hot coffee or snatch at a sharp knife.

Offer a variety of foods at each meal, some foods you know your toddler and baby will like and foods that you like too so that everyone is happy. Offer very tiny portions of foods that are new or ones that your child is not very keen on and encourage your child to have at least one mouthful.

Encourage baby to feed himself with easy-to-pick-up finger foods. This gives the adults a chance to eat their food too. Don't worry about the mess until you get to the end of the meal.

For added protection when your toddler is eating at the table, use a wipe-clean place mat and cover the seat of a dining chair with a towel or teatowel. Better still buy a thin padded seat cover that is removable and machine washable but will tie on to the seat securely.

Children cannot bear to eat foods that are too hot, not just because they may burn their mouths, but because of the frustration. Not being able to eat when the food is there and

least once a day, even if you don't all eat the same food, so that your toddler can learn how to behave by watching the rest of the family. Explain to other members of the family that you are all setting the baby an example – a great way to make everyone pull their socks up, whatever their age.

It is never too early to encourage children to help: laying the table can become quite a game, especially if dolly or teddy has a place set too. Passing plates, bread and salt and pepper can be perilous to begin with but with a little practice becomes second nature. Transform a dull wet Monday tea into a special occasion with a few flowers or candles (well out of reach of very tiny children) and pretty table mats and paper napkins for added decoration.

Try to avoid using a tablecloth with very young children as they can pull the cloth and everything with it on to the floor and themselves, with appalling consequences if there's a hot teapot or hot casserole dish.

Above: *Laying the table is another task that children delight in, and another way to make them feel part of the family mealtime.*

Below: *Even if the child is too young to have exactly the same food, sharing the table is important.*

they are hungry can lead to great outbursts of temper. Many children actually prefer their food to be just lukewarm. Cool broccoli quickly by rinsing with a little cold water. Spoon hot casseroles on to a large plate so they cool quickly and always test the temperature of foods before serving to a child.

TABLE MANNERS

Toddlers can be incredibly messy, but try not to be too fussy and just wipe sticky hands and faces at the end of the meal. Enjoyment is the key. If your child is tucking into her meal with enthusiasm rather than style, then that is the most important part at the beginning – table manners will come as your child becomes more experienced and adept with a knife and fork.

Be understanding and flexible, small children have a lot to learn when they begin to join in with family meals. An active toddler has to learn how to sit still – quite an art in itself – how to hold a knife and fork and how to drink from a cup

Below and below right: *Never forget that food and eating are meant to be fun – don't discipline the child just for the sake of it, and don't impose adult rules too quickly.*

and somehow watch what everyone else is doing and join in too.

The baby has to adapt to sitting up to the table in his highchair and waiting for an adult to offer him a spoonful of food. Not surprisingly there can be a few tantrums and accidents, not to mention mess. Try not to worry. Offer a few finger foods after a bowl of minced foods so that the baby can feed himself, giving you a chance to finish your own meal and tidy up.

Above: *Good manners and sharing provide the perfect atmosphere for good eating – and will aid digestion, too!*

Left: Unusual and varied environments, and the different types of food that go with them, provide huge additional pleasure, and will also help to familiarize your child with situations they will encounter later.

TIPS

- Offer cubes of cheese at the end of a meal to help counteract the potentially harmful effects of any sugary foods eaten.
- Try to encourage children to eat more fruit. Cutting it into small pieces and arranging it on a plate can prove very inviting.
- Encourage children of all ages to drink full-fat milk, either cold, warm or flavoured, as it is a valuable source of the fat-soluble vitamins A and D and the important mineral calcium, vital for healthy bones and teeth.
- If you reheat baby food in the microwave, make sure you stir it thoroughly and leave it to stand for a couple of minutes so any hot spots can even out. Always check the temperature before serving.

- Always make sure a baby is well strapped into a highchair and never leave children unattended while eating in case of accidents.
- Children love the novelty of eating somewhere different. If it's a nice day, why not have a picnic in the local park. If eating outside, make food up separately.
- Snacks can play a vitally important part in a young child's diet as their growth rate is so high it can be difficult to provide sufficient calories and protein at mealtimes alone.
- Make sure you co-ordinate main meals and snacks so that you serve different foods and so that the snacks won't take the edge off the child's appetite for the main meal of the day.

- Try to avoid sweets or cakes and fatty crisps.
- Offer slices of fresh fruit, a few raisins or dried apricots, squares of cheese, a fruit yogurt or milky drink, marmite or smooth peanut butter on toast.
- Make food fun and allow your child to choose where they eat the snack, e.g. in the camp in the garden or at the swings.

Choosing a Balanced and Varied Diet

FOR A BABY, AGED 9–18 MONTHS

Once your baby progresses to more varied puréed meals, you are really beginning to lay down the foundations for a healthy eating pattern which will take your baby through childhood.

It is vitally important to include portions of food from each of the four main food groups shown here per day. But do make sure that the types of food you choose are suitable for the age of your baby.

FOR A TODDLER

Give your child a selection of foods in the four main food groups daily:

Cereal and filler foods: include three to four helpings of the following per day – breakfast cereals, bread, pasta, potatoes, rice.

Fruit and vegetables: try to include at least three or four helpings per day. Choose from fresh, canned, frozen or dried.

Meat and/or alternatives: one to two portions per day – meat – all kinds, including burgers and sausages, poultry, fish (fresh, canned or frozen), eggs (well-cooked), lentils and pulses (for example baked beans, red kidney beans, chick-peas), finely chopped nuts, smooth peanut butter, seeds, tofu, and Quorn.

Dairy foods: include 600ml/1 pint of milk per day or a mix of milk, cheese, yogurt and fromage frais. For a child who goes off drinking milk, try flavouring it or using it in custards, ice cream, rice pudding or cheese sauce. A carton of yogurt or 40g/1½oz of cheese have the same amount of calcium as 190ml/⅓ pint of milk.

Above: *Cereal and filler foods, like bread, potatoes, pasta and rice.*

Above: *Fruit and vegetables, including frozen, dried and canned goods for a child.*

Above: *Meat and meat alternatives, like pulses and nuts (finely ground for a baby).*

Above: *Dairy foods such as milk, cheese and yogurt (plain for a baby, at first).*

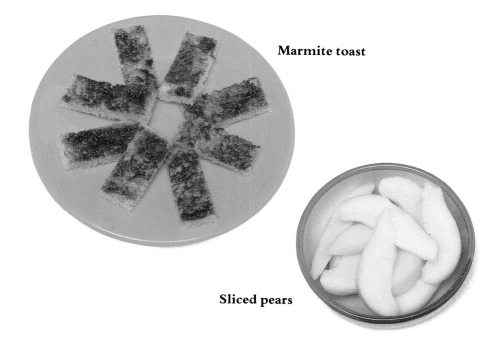

Marmite toast

Sliced pears

FATS

As adults we are all aware of the need to cut down on our fat consumption, but when eating together as a family, bear in mind that fat is a useful source of energy in a child's diet. The energy from fat is in concentrated form, so that your child can take in the calories she needs for growth and development before her stomach becomes overfull. Fat in food is also a valuable source of the fat-soluble vitamins, A, D, E and K, as well as essential fatty acids that the body cannot make by itself.

In general, fat is best provided by foods which contain not just fat but other essential nutrients as well, such as dairy products, eggs, meat and fish. Whole milk and its products such as cheese and yogurt, and eggs contain the fat-soluble vitamins A and D, while sunflower oil, ground nuts and oily fish are a good source of various essential fatty acids.

It is wise to cut down on deep frying and to grill or oven bake foods where possible. All children love chips and crisps but do keep them as a treat rather than a daily snack.

Above: *A good mixture of the four basic food types will provide maximum energy and vitality for growing children.*

FRUIT AND VEGETABLES

Fresh fruit and vegetables play an essential part in a balanced diet. Offer fresh fruit, such as slices of apple or banana, for breakfast and tea, and perhaps thin sticks of carrot and celery for lunch. Instead of biscuits and crisps, offer your child raisins, ready-to-eat apricots, satsumas or carrots and apple slices if she wants a mid-morning or afternoon snack. Keep the fruit bowl within easy reach so your child may be tempted to pick up a banana as she walks through the kitchen.

SNACKS

Young children cannot eat enough food at meal times to meet their needs for energy and growth and snacks can play a vital part in meeting these needs. However keep chocolate biscuits and crisps as a treat. They contain little goodness and are bad for the teeth. At meal times keep sweets out of sight until the main course has been eaten.

Above: *Do keep sweets and chocolate as treats – give fruit and vegetables as day-to-day snacks.*

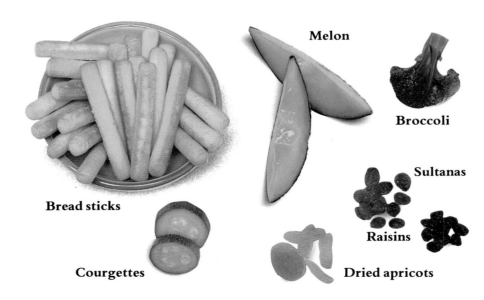

Melon

Broccoli

Sultanas

Raisins

Dried apricots

Bread sticks

Courgettes

Coping With a Fussy Eater

We all have different sized appetites whatever our age, and young children are no exception. Children's appetites fluctuate greatly and often tail off just before a growth spurt. All children go through food fads; some just seem to last longer and be more difficult than others.

A toddler's appetite varies enormously and you may find that she will eat very well one day and eat hardly anything the next. Be guided by your toddler and try to think in terms of what the child has eaten over several days rather than just concentrating on one day.

At the time, it can be really frustrating and worrying. Try not to think of the food that you have just thrown away but try to think more in the long term. Jot down the foods that your child has actually eaten over three or four days or up to a week. You may actually be surprised, it's not just yogurts and crisps after all!

Once you have a list you may find a link between the foods your child eats and the time of day. Perhaps your child eats better when eating with the family, or when the house is quiet. If you do find a link then build on it. You might find that your child is snacking on chocolate, doughnuts, soft drinks or chips when out with friends, and that fussiness at home is really a full tummy. Or it may be that by cutting out a milky drink and a biscuit mid-morning and offering a sliced apple instead, your child may not be so full up come lunch time. Perhaps you could hide the biscuit tin once visitors have had one, so that tiny hands can't keep reaching for more.

If your toddler seems hungrier at breakfast then you could offer eggy bread, a grilled sausage or a few banana slices with her cereal.

Above: *Don't panic about food rejection. Be patient and keep a journal listing what your child actually does eat.*

Right: *Fresh, healthy snacks will preserve the appetite for main meals.*

Although this all sounds obvious, when rushing about caring for a toddler and perhaps an older child or new baby, life becomes rather blurred and it can be difficult to stand back and look at things objectively.

REFUSING TO EAT
A child will always eat if she is hungry although it may not be when you want her to eat. A child can stay fit and healthy on surprisingly little. Providing your child is growing and gaining weight then don't fuss, but if you are worried, talk to your health visitor or doctor. Take the lead from your child, never force feed a child and try not to let meal times become a battle ground.

MAKING MEAL TIMES FUN

Coping with a fussy eater can be incredibly frustrating. The less she eats the crosser you get and so the spiral goes on as your toddler learns how to control meal times. To break this vicious circle, try diffusing things by involving your child in the preparation of the meal. You could pack up a picnic with your child's help, choosing together what to take. Then go somewhere different to eat, it could be the back garden, the swings or even the car. Alternatively, have a dollies' or teddies' tea party or make a camp under the dining table or even in the cupboard under the stairs.

Even very young children enjoy having friends for tea. If your child is going through a fussy or non-eating stage, invite over a little friend with a good appetite. Try to take a back seat and don't make a fuss over how much the visiting child eats.

Above: *Changing the scene and breaking routine can help a lot.*

Below: *Making the meal a special event can distract the child from any eating worries.*

Above: *Getting your child to help you cook the food will encourage them to eat it, too.*

Above: *Children are more likely to eat with friends of their own age around them.*

10 TIPS TO COPE WITH A FUSSY EATER

1 Try to find meals that the rest of the family enjoys and where there are at least one or two things the fussy child will eat as well. It may seem easier to cook only foods that your child will eat but it means a very limited diet for everyone else and your child will never get the chance to have a change of mind and try something new.

2 Serve smaller portions of food to your child.

3 Invite round her friend with a hearty appetite. A good example sometimes works but don't comment on how much the visiting child has eaten.

4 Invite an adult who the child likes for supper – a granny, uncle or friend. Sometimes a child will eat for someone else without any fuss at all.

5 Never force feed a child.

6 If your child is just playing with the food and won't eat, quietly remove the plate without fuss and don't offer pudding.

7 Try to make meal times enjoyable and talk about what the family has been doing.

8 Try limiting snacks and drinks between meals so your child feels hungrier when it comes to family meal times. Alternatively, offer more nutritious snacks and smaller main meals if your child eats better that way.

9 Offer drinks after a meal so that they don't spoil the appetite.

10 Offer new foods when you know your child is hungry and hopefully more receptive.

Above: *Remember to give drinks after the meal, not before.*

EATING TOGETHER

Eating together as a family should be a happy part of the day, but can turn into a nightmare if everyone is tired or you feel as though the only things your children will eat are chips.

There is nothing worse than preparing a lovely supper, laying the table and sitting down with everyone and then one child refuses to eat, shrieks her disapproval or just pushes the food around the plate. However hard you try to ignore this behaviour, the meal is spoilt for everyone, especially if this is a regular occurrence. It's not fair on you or anyone else.

If you feel this is just a passing phase, then you could try just ignoring it and carry on regardless. Try to praise the good things, perhaps the way the child sits nicely at the table or the way she holds a knife and fork. Talk about the things that have been happening in the day, rather than concentrating on the meal itself. Try to avoid comparing your child's appetite with more hearty eaters. With luck, this particular fad will go away.

However, if it becomes a regular thing and mealtimes always seem more like a battleground than a happy family gathering, perhaps it's time for a sterner approach.

First steps

● Check to see if there is something physically wrong with your child. Has she been ill? If she has, she may not have recovered fully. If you're worried, then ask your doctor.

● Perhaps your child has enlarged adenoids or tonsils which could make swallowing difficult, or perhaps she has a food allergy, such as coeliacs disease – an intolerance to gluten – which may be undiagnosed but which would give the child tummy pains after eating. Again, check with your doctor.

● Is your child worried or stressed? If your family circumstances have changed – a new baby perhaps, or if you've moved recently – your child may be unhappy or confused.

● Is your child trying to get your attention?

Above: *Good seating of the right height will contribute to comfort and relaxation.*

Secondly

Look at the way in which you as a family eat. Do you eat at regular times? Do you sit down to eat or catch snacks on the move? Do you enjoy your food or do you always feel rushed and harassed? Children will pick up habits from their parents – bad as well as good. If you don't tend to sit down to a meal or you have the habit of getting up during meal times to do other jobs, then it's hard to expect the child to behave differently.

Finally

Talk things over with all the family. If you all feel enough is enough, then it's time to make a plan of action. Explain that from now on you are all going to eat together where possible, when and where you say so. You will choose the food, there will be three meals a day and no snacks. Since milk is filling and dulls the appetite, milky drinks will only be given after a meal; during the meal water or juice will be provided.

It is vitally important to involve all the family in this strategy so that there is no dipping into the biscuit tin or raiding the cupboard for crisps after school. Make sure that the fussy eater is aware of what is going to happen and give a few days' notice so that the idea can sink in.

Once you have outlined your strategy, work out your menus and

stick to them. Include foods your child definitely likes, chicken or carrots for instance, and obviously avoid foods your child hates although you could introduce some new foods for variety. Set yourself a time scale, perhaps one or two weeks, and review things after this period has elapsed.

PUTTING THE PLAN INTO ACTION

Begin your new plan of action when all the family are there to help, such as a weekend, and stick to it. Make a fuss of the plans so it seems more like a game than a prison sentence. Add a few flowers to the table or a pretty cloth to make it more special.

Begin the day with a normal breakfast but give the fussy eater the smallest possible portion. If the child eats it up then offer something you know your child likes, such as an apple, a few raisins or a fruit yogurt.

As the days progress, you could offer a biscuit or milkshake as a treat.

Give plenty of encouragement and praise, but be firm if the child plays up. If she behaves badly, take her to a different room or to the bottom of the stairs and explain that the only food is that on the table. Sit down with the rest of the family, leaving the fussy eater's food on the table and try to ignore the child.

If the child changes her mind just as you're about to clear away, then get the other members of the family to come back to the table and wait until the fussy eater has finished.

Continue in this way with other meals. Don't be swayed if your child says she will eat her food watching TV or if she wants her pudding first. Explain that she must eat just like everyone else or go without.

If she begins to cry, sit her down in another room and return to the table. This is perhaps the hardest thing of all.

After a few days, there should be a glimmer of progress. Still offer tiny portions of food, followed by foods that you know your child will eat as a treat. Keeping to a plan like this is hard, but if all the family sticks together and thinks positively, then it is possible. Keep to the time span you decided, then suggest you all go to your local pizza or burger restaurant and let the fussy eater choose what she likes.

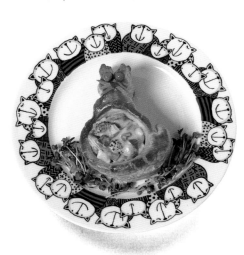

MEATY MAIN MEALS

TRYING TO PLEASE ALL THE FAMILY ALL OF THE TIME CAN BE A BIT OF A TALL ORDER, ESPECIALLY WHEN TRYING TO COOK TASTY MEALS ON A BUDGET. TRY THESE DELICIOUS NEW WAYS WITH QUICK-TO-COOK MINCE AND CHOPS, PLUS A SPEEDY STIR FRY, OR SLOW-COOK BEEF BOURGIGNON, OSSO BUCCO PORK OR LAMB HOT-POT.

Mediterranean Lamb

3 lamb chump chops

350g/12oz courgettes

½ yellow pepper

½ red pepper

3 tomatoes

1 garlic clove, crushed

15ml/1 tbsp clear honey

few sprigs of fresh rosemary, plus extra to garnish

15ml/1 tbsp olive oil

200g/7oz can baked beans

salt and freshly ground black pepper

crusty bread, to serve

2 Season two of the chops for the adults with garlic, honey, rosemary and salt and pepper, and drizzle oil over the vegetables.

3 Cook under a hot grill for 12–14 minutes, turning once, until the lamb is well browned and cooked through and the vegetables are browned.

5 Chop or process the remaining lamb with four slices of courgette, two small pieces of pepper, two peeled tomato quarters and 15–30ml/1–2 tbsp of baked beans, adding a little boiled water if too dry. Spoon into a dish for the baby and test the temperature of the children's food before serving.

6 For the adult's portions, discard the cooked rosemary. Spoon the pan juices over the chops and garnish with fresh rosemary. Serve with crusty bread.

1 Rinse the chops under cold water, pat dry, trim off fat and place in the base of a grill pan. Trim and slice the courgettes, cut away the core and seeds from the peppers and then rinse and cut into chunky pieces. Rinse and cut the tomatoes into quarters and arrange the vegetables around the lamb.

4 Warm the baked beans in a small saucepan. Drain and transfer the unseasoned chop to a chopping board and cut away the bone and fat. Thinly slice half the meat for the toddler and arrange on a plate with a few of the vegetables and 30–45ml/ 2–3 tbsp of the baked beans.

TIP
Lamb varies hugely in price throughout the year. Depending on the season and price you may prefer to use lamb cutlets or loin chops. Allow two chops per adult and reduce the cooking time slightly as these chops are smaller. Do not give honey to young children. Use redcurrant jelly in place of honey for baby and adults if preferred.

Lamb Hot-pot

350g/12oz lamb fillet

1 onion

1 carrot

175g/6oz swede

15ml/1 tbsp sunflower oil

30ml/2 tbsp plain flour

450ml/¾ pint/1⅞ cups lamb stock

15ml/1 tbsp fresh sage or 1.5ml/
¼ tsp dried

50g/2oz black pudding (optional)

½ dessert apple

275g/10oz potatoes

15g/½oz/1 tbsp butter

225g/8oz Brussels sprouts

salt and freshly ground black pepper

1 Preheat the oven to 180°C/ 350°F/Gas 4. Rinse the lamb under cold water, pat dry, trim away any fat and then slice thinly. Peel and chop the onion, carrot and swede.

2 Heat the oil in a large frying pan and fry the lamb, turning until browned on both sides. Lift the lamb out of the pan, draining off any excess oil and transfer one-third to a small 600ml/1 pint/2½ cup casserole dish for the children and the rest to a 1.2 litre/2 pint/5 cup casserole dish for the adults.

3 Add the vegetables to the pan and fry for 5 minutes, stirring until lightly browned.

4 Stir in the flour, then add the stock and sage. Bring to the boil, stirring, then divide between the two casserole dishes.

5 Peel and chop the black pudding if using, core, peel and chop the apple and add both to the larger casserole dish with a little seasoning.

6 Thinly slice the potatoes and arrange overlapping slices over both casserole dishes. Dot with butter and season the larger dish.

7 Cover and cook in the oven for 1¼ hours. For a brown topping, remove the lid and grill for a few minutes at the end of cooking until browned. Cook the Brussels sprouts in boiling water for 8–10 minutes until tender and drain.

8 Chop or process half the hot-pot from the small casserole with a few sprouts for the baby, adding extra gravy if needed until the desired consistency is reached. Spoon into a baby dish.

9 Spoon the remaining child's hotpot on to a plate, add a few sprouts and cut up any large pieces. Test the temperature of the children's food before serving.

10 Spoon the hotpot for the adults on to serving plates and serve with Brussels sprouts.

TIP
Traditionally neck or scrag end of lamb would have been used for making a hotpot. This cut does require very long slow cooking and can be fatty with lots of bones so not ideal for children. Fillet of lamb is very lean, with very little waste and makes a tasty and healthier alternative.

Beef Korma

350g/12oz lean minced beef

1 onion, chopped

1 carrot, chopped

1 garlic clove, crushed

400g/14oz can tomatoes

pinch of dried mixed herbs

25g/1oz pasta shapes

50g/2oz creamed coconut

50g/2oz button mushrooms

50g/2oz fresh spinach leaves

15ml/1 tbsp hot curry paste

salt and freshly ground black pepper

boiled rice, warm naan bread and a
    little grated Cheddar cheese
    (optional), to serve

1 Dry fry the mince and onion in a
medium-sized saucepan, stirring
until the beef is browned.

2 Add the carrot, garlic, tomatoes
and herbs, bring to the boil,
stirring and then cover and simmer
for about 30 minutes, stirring
occasionally.

3 Cook the pasta in a small
saucepan of boiling water for 10
minutes until tender. Drain.

4 Meanwhile place the coconut in
a small bowl, add 120ml/4fl oz/
½ cup of boiling water and stir until
dissolved. Wipe and slice
mushrooms and wash and drain
spinach, discarding any large stalks.

5 Transfer one-third of the mince
to another saucepan. Stir the
coconut mixture, curry paste and
salt and pepper into the remaining
mince and cook for about 5 minutes,
stirring.

6 Mash or process one-third of the
pasta and reserved mince
mixture to the desired consistency
for baby and spoon into a small dish.

7 Spoon the remaining pasta and
reserved mince into a small bowl
for the older child.

8 Stir the mushrooms and spinach
into the curried mince and cook
for 3–4 minutes until the spinach has
just wilted. Spoon on to warmed
serving plates for the adults and
serve with boiled rice and warmed
naan bread. Sprinkle the toddler's
portion with a little cheese, if liked.
Test the temperature of the
children's food before serving.

TIP
Weigh spinach after it has been
picked over and stalks removed or,
if you're short of time, buy ready
prepared spinach. If you can't get
creamed coconut, use 120ml/
4fl oz/½ cup coconut milk instead,
or soak 25g/1oz/1 cup desiccated
coconut in 150ml/¼ pint/⅔ cup
boiling water for 30 minutes then
strain and use the liquid.

Bobotie with Baked Potatoes and Broccoli

3 medium baking potatoes

1 onion, chopped

350g/12oz lean minced beef

1 garlic clove, crushed

10ml/2 tsp mild curry paste

10ml/2 tsp wine vinegar

90ml/6 tbsp fresh breadcrumbs

15ml/1 tbsp tomato purée

25g/1oz sultanas

15ml/1 tbsp mango chutney

1 medium banana, sliced

2 eggs

20ml/4 tsp turmeric

120ml/4fl oz/½ cup skimmed milk

4 small bay leaves

225g/8oz broccoli, cut into florets

30ml/2 tbsp fromage frais or Greek yogurt

200g/7oz can baked beans

salt and freshly ground black pepper

1 Preheat the oven to 180°C/ 350°F/Gas 4. Scrub the potatoes, insert a skewer into each and bake for 1½ hours until tender.

2 Place the chopped onion and 225g/8oz of the beef in a saucepan and dry fry, until browned all over, stirring frequently.

3 Add the garlic and curry paste, stir well and cook for 1 minute, then remove from the heat and stir in the vinegar, 60ml/4 tbsp of the breadcrumbs, the tomato purée, sultanas and a little salt and pepper.

TIP
Serve any leftover adult portions cold with salad.

4 Chop up any large pieces of mango chutney and stir into the meat mixture with the banana slices. Spoon into a 900ml/1½ pint/3¾ cup ovenproof dish and press into an even layer with the back of a spoon.

5 Place the dish on a baking sheet, cover loosely with foil and cook for 20 minutes.

6 Meanwhile mix the remaining beef with the remaining breadcrumbs, then beat the eggs together and stir 15ml/1 tbsp into the meat. Make eight small meatballs about the size of a grape for the baby. Form the remaining beef into a 7.5cm/3in burger using an upturned biscuit cutter as a mould.

7 Blend the turmeric, milk and a little salt and pepper with the remaining eggs. Remove the cover from the Bobotie, and lay the bay leaves over the meat.

8 Pour the egg mixture over. Return to the oven for a further 30 minutes until well risen and set.

9 When the adults' portion is ready, heat the grill and cook the burger and meatballs until browned, turning once. The burger will take 8–10 minutes, while the meatballs will take about 5 minutes.

10 Cook the broccoli in boiling water until tender and drain.

11 Cut the Bobotie for the adults into wedges and serve with jacket potatoes topped with fromage frais and broccoli.

12 Serve the toddler's burger with half a potato, warmed baked beans and a few broccoli florets. Serve the baby's meatballs with chunky pieces of peeled potato and broccoli. Spoon a few baked beans into a small dish for baby. Test the temperature of the food before serving to the children.

Moussaka

1 onion, chopped

350g/12oz minced lamb

400g/14oz can tomatoes

1 bay leaf

1 medium aubergine, sliced

2 medium potatoes

1 medium courgette, sliced

30ml/2 tbsp olive oil

2.5ml/½ tsp grated nutmeg

2.5ml/½ tsp ground cinnamon

2 garlic cloves, crushed

salt and freshly ground black pepper

For the Sauce

30ml/2 tbsp sunflower margarine

30ml/2 tbsp plain flour

200ml/7fl oz/1 cup milk

pinch of grated nutmeg

15ml/1 tbsp freshly grated Parmesan cheese

20ml/4 tsp fresh breadcrumbs

1 Dry fry the onion and minced lamb in a saucepan until browned, stirring. Add the tomatoes and bay leaf, bring to the boil, stirring, then cover and simmer for 30 minutes.

2 Place the aubergine slices in a single layer on a baking sheet, sprinkle with a little salt and set aside for 20 minutes. Preheat oven to 200°C/400°F/Gas 6.

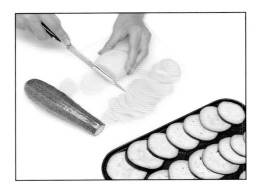

3 Slice the potatoes thinly and cook in boiling water for 3 minutes. Add the courgette and cook for 2 minutes until tender.

4 Remove most of the slices with a slotted spoon and place in a colander, leaving just enough for the baby portion. Cook these for 2–3 minutes more until soft, then drain. Rinse the vegetables and drain well.

5 Rinse the salt off the aubergine and pat dry. Heat the oil in a frying pan and fry the aubergines until browned on both sides. Drain.

6 Spoon 45ml/3 tbsp of the mince mixture into a bowl with the baby vegetables and chop or purée to the desired consistency.

7 Spoon 60ml/4 tbsp of the mince mixture into an ovenproof dish for the older child with four slices of potato overlapping over the top, then add a slice of aubergine and three slices of courgette.

8 Stir the spices, garlic and seasoning into the remaining lamb mixture, and cook for 1 minute and then spoon into a 1.2 litre/ 2 pint/5 cup shallow ovenproof dish discarding the bay leaf.

9 Arrange the remaining potatoes overlapping on top of the lamb, and then add the aubergine slices, tucking the courgette slices in between the aubergines in a random pattern.

10 To make the sauce, melt the margarine in a small saucepan, stir in the flour then gradually add the milk and bring to the boil, stirring until thickened and smooth. Add a pinch of nutmeg and a little salt and pepper.

11 Pour a little of the sauce over the toddler's portion then pour the rest over the adults' portion. Sprinkle the larger dish with Parmesan cheese and 15ml/3 tsp breadcrumbs, sprinkling the remaining breadcrumbs over the toddler's portion.

12 Cook the moussakas in the oven. The larger dish will take 45 minutes while the toddler's portion will take about 25 minutes. Reheat the baby portion, and test the temperature of the children's food before serving.

Chilli con Carne

3 medium baking potatoes

1 onion, chopped

450g/1lb lean minced beef

1 carrot, chopped

½ red pepper, cored, seeded and diced

400g/14oz can tomatoes

10ml/2 tsp tomato purée

150ml/¼ pint/⅔ cup beef stock

3 small bay leaves

30ml/2 tbsp olive oil

115g/4oz button mushrooms, sliced

2 garlic cloves, crushed

10ml/2 tsp mild chilli powder

2.5ml/½ tsp ground cumin

5ml/1 tsp ground coriander

200g/7oz can red kidney beans, drained

40g/1½oz frozen mixed vegetables

15ml/1 tbsp milk

knob of butter or margarine

60ml/4 tbsp fromage frais or Greek yogurt

15ml/1 tbsp chopped fresh coriander leaves

salt and freshly ground black pepper

green salad, to serve

1 Preheat the oven to 180°C/ 350°F/Gas 4. Scrub and prick the potatoes and cook in the oven for 1½ hours. Dry fry the onion and minced beef in a saucepan until browned. Add the carrot and red pepper and cook for 2 minutes.

2 Add the tomatoes, tomato purée and stock and bring to the boil. Transfer one quarter of the mixture to a 600ml/1 pint/2½ cup casserole dish, add 1 of the bay leaves, cover with a lid and set aside.

3 Spoon the remaining mince mixture into a 1.2 litre/2 pint/ 5 cup casserole. Heat 15ml/1 tbsp of the oil in the same pan and fry the mushrooms and garlic for 3 minutes.

4 Stir in the spices and seasoning, cook for 1 minute then add the drained red kidney beans and the remaining bay leaves and stir into the meat mixture. Cover and cook both dishes in the oven for 1 hour.

5 When the potatoes are cooked, cut into halves or quarters and scoop out the potato leaving a thin layer of potato on the skin.

6 Brush the potato skins with the remaining oil and grill for 10 minutes until browned.

7 Boil the frozen mixed vegetables for 5 minutes and mash the potato centres with milk and a knob of butter or margarine.

8 Spoon the mince from the smaller casserole into an ovenproof ramekin dish for the toddler and the rest into a bowl for baby. Top both with some of the mashed potato.

9 Drain the vegetables, arrange two pea eyes, carrot pieces for the mouth and mixed vegetables for hair on the toddler's dish.

10 Spoon the remaining vegetables into a baby bowl and chop or process to the desired consistency. Test the temperature of the children's food before serving.

11 Spoon the adults' chilli on to warmed serving plates, add the potato skins and top with fromage frais and chopped coriander. Serve with a green salad.

Beef Bourguignon with Creamed Potatoes

450g/1lb stewing beef

15ml/1 tbsp sunflower oil

1 onion, chopped

30ml/2 tbsp plain flour

300ml/½ pint/1¼ cups beef stock

15ml/1 tbsp tomato purée

small bunch of fresh herbs or 1.5ml/¼ tsp dried

2 garlic cloves, crushed

90ml/6 tbsp red wine

75g/3oz shallots

75g/3oz button mushrooms

25g/1oz/2 tbsp butter

450g/1lb potatoes

175g/6oz green cabbage

30–60ml/2–4 tbsp milk

salt and freshly ground black pepper

a few fresh herbs, to garnish

1 Preheat the oven to 160°C/325°F/Gas 3. Trim away any fat from the beef and cut into cubes.

2 Heat the oil in a frying pan, add half of the beef and fry until browned. Transfer to a plate and fry the remaining beef and the chopped onion until browned.

3 Return the first batch of beef to the pan with any meat juices, stir in the flour and then add the stock and tomato purée. Bring to the boil, stirring, until thickened.

4 Spoon one-third of the beef mixture into a small 600ml/½ pint/1¼ cup casserole dish for the children, making sure that the meat is well covered with stock. Add a few fresh herbs or half a dried bay leaf. Set the casserole aside.

5 Add the remaining herbs, garlic, wine and seasoning to the beef in the frying pan and bring to the boil. Transfer the beef to a 1.2 litre/2 pint/5 cup casserole dish. Cover both dishes and cook in the oven for 2 hours or until the meat is very tender.

6 Meanwhile, halve the shallots if large, wipe and slice the mushrooms, cover and put to one side.

7 Half an hour before the end of cooking, fry the shallots in a little butter until browned, then add the mushrooms and fry for 2–3 minutes. Stir into the larger casserole and cook for the remaining time.

8 Cut the potatoes into chunky pieces and cook in a saucepan of boiling, salted water for 20 minutes. Shred the cabbage, discarding the hard core, rinse and steam above the potatoes for the last 5 minutes.

9 Drain the potatoes and mash with 30ml/2 tbsp of the milk and the remaining butter.

10 Chop or process one-third of the child's casserole with a spoonful of cabbage, adding extra milk if necessary. Spoon into a dish for the baby, with a little potato.

11 Spoon the remaining child's casserole on to a plate for the toddler. Remove any bay leaf and cut up any large pieces of beef. Add potato and cabbage.

12 Garnish the adults' casserole with fresh herbs and serve with potatoes and cabbage. Test the temperature of the children's food.

Osso Bucco Pork with Rice

3 pork spare rib chops, about 500g/
  1¼lb

15ml/1 tbsp olive oil

1 onion, chopped

1 large carrot, chopped

2 celery sticks, thinly sliced

2 garlic cloves, crushed

400g/14oz can tomatoes

few sprigs fresh thyme or 1.5ml/
  ¼ tsp dried

grated rind and juice of ½ lemon

150g/5oz/⅔ cup long grain rice

pinch of turmeric

115g/4oz green beans

knob of butter

20ml/4 tbsp freshly grated Parmesan
  cheese

salt and freshly ground black pepper

1 or 2 sprigs parsley

1 Preheat the oven to 180°C/
350°F/Gas 4. Rinse the chops
under cold water and pat dry.

2 Heat the oil in a large frying pan,
add the pork, brown on both
sides and transfer to a casserole dish.

3 Add the onion, carrot and celery
to the frying pan and fry for 3
minutes until lightly browned.

4 Add half the garlic, the
tomatoes, thyme and lemon
juice and bring to the boil, stirring.
Pour the mixture over the pork,
cover and cook in the oven for 1½
hours or until tender.

5 Half fill two small saucepans
with water and bring to the boil.
Add 115g/4oz/½ cup rice to one
with a pinch of turmeric and salt,
add the remaining rice to the second
pan. Return to the boil, and simmer.

6 Trim the beans and steam above
the larger pan of rice for 8
minutes or until the rice is tender.

7 Drain both pans of rice, and
return the yellow rice to the pan
and add the butter, Parmesan cheese
and a little pepper. Mix together
well and keep warm.

8 Dice one chop, discarding any
bone if necessary. Spoon a little
of the white rice on to a plate for the
toddler and add half of the diced
chop. Add a few vegetables and a
spoonful of the sauce. Process the
other half of chop, some vegetables,
rice and sauce to the desired
consistency and turn into a small
bowl for the baby.

9 Spoon the yellow rice on to the
adults' plates and add a pork
chop to each. Season the sauce to
taste and then spoon the sauce and
vegetables over the meat, discarding
the thyme sprigs if used. Finely chop
the parsley and sprinkle over the
pork with the lemon rind and the
remaining crushed garlic.

10 Serve the adult's and
toddler's portions with green
beans. Test the temperature of the
children's food before serving.

Pork Stir Fry

250g/9oz pork fillet

1 courgette

1 carrot

½ green pepper

½ red pepper

½ yellow pepper

200g/7oz packet fresh beansprouts

20ml/4 tsp sunflower oil

25g/1oz cashew nuts

150ml/¼ pint/⅔ cup chicken stock

30ml/2 tbsp tomato ketchup

10ml/2 tsp cornflour, blended with
15ml/1 tbsp cold water

1 garlic clove, crushed

10ml/2 tsp soy sauce

20ml/4 tsp yellow bean paste or
Hoi-Sin sauce

1 Rinse the pork under cold water, cut away any fat and thinly slice. Halve the courgette lengthways and then slice thinly, thinly slice the carrot, and cut the peppers into thin strips, discarding the core and the seeds. Place the beansprouts in a sieve and rinse well.

2 Heat 15ml/3 tsp of the oil in a wok or large frying pan, add the cashews and fry for 2 minutes until browned. Drain and reserve.

3 Add the sliced pork and stir fry for 5 minutes until browned all over and cooked through. Drain and keep warm.

4 Add the remaining oil and stir fry the courgette and carrot for 2 minutes. Add the peppers and fry for a further 2 minutes.

5 Stir in the beansprouts, stock, ketchup and the cornflour mixture, bring to the boil, stirring until the sauce has thickened.

6 Transfer 1 large spoonful of the vegetables to a bowl and 2–3 large spoonfuls to a plate for the children and set aside. Add the garlic and soy sauce to the pan and cook for 1 minute.

7 Chop or process four slices of pork with the baby's reserved vegetables to the desired consistency and turn into a baby bowl. Arrange six slices of pork on the child's plate with the reserved vegetables. Spoon the vegetables in the wok on to two adult serving plates.

8 Return the remaining pork to the wok with the nuts and yellow bean paste or Hoi-Sin sauce and cook for 1 minute, spoon on to the adults' plates and serve immediately. Test the temperature of the children's food before serving.

Sausage Casserole

450g/1lb large sausages

15ml/1 tbsp sunflower oil

1 onion, chopped

225g/8oz carrots, chopped

400g/14oz can mixed beans in water, drained

15ml/1 tbsp plain flour

450ml/¾ pint/1⅞ cup beef stock

15ml/1 tbsp Worcestershire sauce

15ml/3 tsp tomato purée

15ml/3 tsp soft brown sugar

10ml/2 tsp Dijon mustard

1 bay leaf

1 dried chilli, chopped

3 medium baking potatoes

salt and freshly ground black pepper

butter and sprigs of fresh parsley, to serve

1 Preheat the oven to 180°C/ 350°F/Gas 4. Prick and separate the sausages.

2 Heat the oil in a frying pan, add the sausages and cook over a high heat until evenly browned but not cooked through. Drain and transfer to a plate.

3 Add the chopped onion and carrots to the pan and fry until lightly browned. Add the drained beans and flour, stir well and then spoon one-third of the mixture into a small casserole. Stir in 150ml/ ¼ pint/⅔ cup stock, 5ml/1 tsp tomato purée and 5ml/1 tsp sugar.

4 Add the Worcestershire sauce, remaining beef stock, tomato purée and sugar to the pan, together with the mustard, bay leaf and chopped chilli. Season and bring to the boil, then pour the mixture into a large casserole.

5 Add two sausages to the small casserole and the rest to the larger dish. Cover both and cook in the oven for 1½ hours.

6 Scrub and prick the potatoes and cook on a shelf above the casserole for 1½ hours until tender.

7 Spoon two-thirds of the child's casserole on to a plate for the toddler. Slice the sausages and give him or her two-thirds. Halve one of the baked potatoes, add a knob of butter and place on the toddler's plate.

8 Scoop the potato from the other half and mash or process with the remaining child's beans to the desired consistency for baby. Spoon into a dish and serve the remaining sausage slices as finger food.

9 Spoon the adults' casserole on to plates, halve the other potatoes, add butter and garnish with parsley. Test the temperature of the children's food.

PERFECT POULTRY

WHATEVER THE AGE OF THE DINER, CHICKEN AND TURKEY ARE ALWAYS POPULAR. AVAILABLE IN A RANGE OF CUTS AND PRICES THERE'S SOMETHING TO SUIT ALL BUDGETS AND TASTES FROM A TASTY SALAD TO PAN FRIED TURKEY, TANDOORI STYLE MARINADE OR STUFFED CHICKEN BREASTS WITH PESTO, HAM AND CHEESE.

Chicken and Thyme Casserole

6 chicken thighs

15ml/1 tbsp olive oil

1 onion, chopped

30ml/2 tbsp plain flour

300ml/½ pint/1¼ cups chicken stock

few sprigs fresh thyme or 1.5 ml/ ¼ tsp dried

To Serve

350g/12oz new potatoes

1 large carrot

115g/4oz/¾ cup frozen peas

5ml/1 tsp Dijon mustard

grated rind and juice of ½ orange

60ml/4 tbsp fromage frais or natural yogurt

salt and freshly ground black pepper

fresh thyme or parsley, to garnish

1 Preheat the oven to 180°C/ 350°F/Gas 4. Rinse the chicken under cold water and pat dry.

2 Heat the oil in a large frying pan, add the chicken and brown on both sides, then transfer to a casserole.

3 Add the onion and fry, stirring until lightly browned. Stir in the flour, then add the stock and thyme and bring to the boil, stirring.

4 Pour over the chicken, cover and cook in the oven for 1 hour or until tender.

5 Meanwhile scrub the potatoes and cut any large ones in half. Cut the carrot into matchsticks.

6 Cook the potatoes in boiling water 15 minutes before the chicken is ready and cook the carrots and peas in a separate pan of boiling water for 5 minutes. Drain.

7 Take one chicken thigh out of the casserole, remove the skin and cut the meat away from the bone. Place in a baby bowl or a food processor or blender with some of the vegetables and gravy and chop or process to the desired consistency. Turn into a baby bowl.

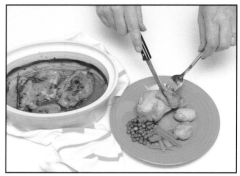

8 Take a second chicken thigh out of the casserole for the toddler, remove the skin and bone and slice if necessary. Arrange on a plate with some of the vegetables and gravy. Test the temperature of the children's food before serving.

9 Arrange two chicken thighs on to warmed serving plates for the adults. Stir the mustard, orange rind and juice, fromage frais and seasoning into the hot sauce and then spoon over the chicken. Serve at once, with the vegetables, garnished with a sprig of thyme or parsley.

Tandoori Style Chicken

6 chicken thighs

150g/5oz natural yogurt

6.5ml/1¼ tsp paprika

5ml/1 tsp hot curry paste

5ml/1 tsp coriander seeds, roughly crushed

2.5ml/½ tsp cumin seeds, roughly crushed

2.5ml/½ tsp turmeric

pinch of dried mixed herbs

5ml/2 tsp sunflower oil

To Serve

350g/12oz new potatoes

3 celery sticks

10cm/4in piece cucumber

15ml/1 tbsp olive oil

5ml/1 tsp white wine vinegar

5ml/1 tsp mint jelly

salt and freshly ground black pepper

few sprigs of watercress, knob of butter and cherry tomatoes, to serve

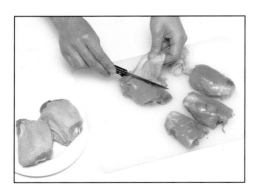

1 Cut away the skin from the chicken thighs and slash the meat two or three times with a small knife. Rinse under cold water and pat dry.

2 Place four thighs in a shallow dish, the other two on a plate. Place the yogurt, 5ml/1 tsp of the paprika, the curry paste, both seeds and almost all the turmeric into a small bowl and mix together. Spoon over the four chicken thighs.

3 Sprinkle the remaining paprika and the mixed herbs over the other chicken thighs and sprinkle the remaining pinch of turmeric over 1 thigh. Cover both dishes loosely with clear film and chill in the refrigerator for 2–3 hours.

4 Preheat the oven to 200°C/ 400°F/Gas 6. Arrange the chicken thighs on a roasting rack set over a small roasting tin and drizzle oil over the herby chicken. Pour a little boiling water into base of the tin and cook for 45–50 minutes or until the juices run clear when the chicken is pierced with a skewer.

5 Meanwhile, scrub the potatoes and halve any large ones. Cook in a saucepan of boiling water for 15 minutes until tender.

6 Cut one celery stick and a small piece of cucumber into matchsticks. Chop or shred the remaining celery and cucumber.

7 Blend together the oil, vinegar, mint jelly and seasoning in a bowl and add the chopped or shredded celery, cucumber and watercress, tossing well to coat.

8 Drain the potatoes and toss in a little butter. Divide the potatoes among the adults' plates, toddler's dish and the baby bowl. Cut the chicken off the bone for the toddler and arrange on a plate with half of the celery and cucumber sticks.

9 Cut the remaining chicken thigh into tiny pieces for the baby, discarding the bone, and allow to cool. Add to the bowl with the cooled potatoes and vegetable sticks and allow baby to feed him- or herself. Add a few halved tomatoes to each portion. Test the temperature of children's food before serving.

10 For the adults, arrange the chicken thighs on warmed serving plates with the potatoes. Serve with the piquant salad.

Chicken Wrappers

3 boneless, skinless chicken breasts

5ml/1 tsp pesto

40g/1½oz wafer thin ham

50g/2oz Cheddar cheese

350g/12oz new potatoes

175g/6oz green beans

15g/½oz/1 tbsp butter

10ml/2 tsp olive oil

1 tomato

6 stoned black olives

10ml/2 tsp plain flour

150ml/¼ pint/⅔ cup chicken stock

15ml/1 tbsp crème fraîche or soured cream (optional)

few sprigs of fresh basil (optional)

1 Rinse the chicken under cold water and pat dry. Put one chicken breast between two pieces of clear film and bat out with a rolling pin until half as big again. Repeat with the other two chicken breasts.

TIP
Depending on the age of the baby, you may prefer to omit the sauce and serve the meal as finger food, cutting it into more manageable pieces first.

2 Spread pesto over two of the chicken breasts and divide the ham among all three. Cut the cheese into three thick slices then add one to each piece of chicken. Roll up so that the cheese is completely enclosed, then secure with string.

3 Scrub the potatoes, halve any large ones and cook in a saucepan of boiling water for 15 minutes until tender. Trim the beans and cook in a separate pan of boiling water for 10 minutes.

4 Meanwhile, heat the butter and oil in a large frying pan, add the chicken and cook for about 10 minutes, turning several times, until well browned and cooked through.

5 Lift the chicken out of the pan, keeping those spread with pesto warm, and leaving the other one to cool slightly.

6 Cut the tomato into wedges and halve the olives. Stir in the flour and cook for 1 minute. Gradually stir in the stock and bring to the boil, stirring until thickened. Add the tomato and olives.

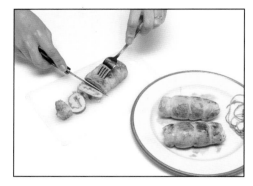

7 Snip the string off the chicken and slice the children's chicken breast thinly. Arrange four chicken slices on a child's plate with a couple of spoonfuls of sauce and a few potatoes and green beans. (Don't offer olives unless the child is a very adventurous eater.)

8 Chop or process the remaining cut chicken with 30ml/2 tbsp of the sauce, one or two potatoes and three green beans to the desired consistency. Test the temperature of the children's food before serving.

9 Arrange the remaining chicken on plates. Add the crème fraîche or soured cream to the pan, if using, and heat gently. Spoon over the chicken and serve with vegetables.

Pan Fried Turkey with Coriander

3 turkey breast steaks

1 onion

1 red pepper, cored and seeded

15ml/1 tbsp sunflower oil

5ml/1 tsp plain flour

150ml/¼ pint/⅔ cup chicken stock

30ml/2 tbsp frozen sweetcorn

150g/5oz/⅔ cup long grain white rice

1 garlic clove

1 dried chilli

50g/2oz creamed coconut

30ml/2 tbsp chopped fresh coriander

fresh coriander sprigs and lime wedges, to serve

1 Rinse the turkey steaks under cold water, and pat dry. Chop one of the steaks, finely chop one quarter of the onion and dice one quarter of the red pepper. Heat 5ml/ 1 tsp of the oil in a small frying pan and fry the diced turkey and chopped onion until lightly browned all over.

2 Stir in the flour, add the stock, sweetcorn and chopped pepper, then bring to the boil, cover and simmer for 10 minutes.

3 Cook the long grain rice in boiling water for 8–10 minutes or until just tender. Drain.

4 Meanwhile process or finely chop the remaining pieces of onion and red pepper with the garlic and chilli, including the seeds if you like hot food.

5 Put the creamed coconut into a bowl, pour on 200 ml/7fl oz/ 1 cup boiling water and stir until the coconut is completely dissolved.

6 Heat the remaining oil in a large frying pan. Brown the turkey breasts on one side, turn over and add the pepper paste. Fry for 5 minutes until the second side of the turkey is also browned.

7 Pour the coconut milk over the turkey and cook for 2–3 minutes, stirring until the sauce has thickened slightly. Sprinkle with the chopped fresh coriander.

8 Chop or process one-third of the children's portion with 30ml/ 2 tbsp of the rice until the desired consistency is reached. Spoon into a baby bowl.

9 Spoon the child's portion on to a child's plate and serve with a little rice. Test the temperature of the children's food before serving.

10 Spoon the rest of the rice on to warmed serving plates for the adults, add the turkey and sauce and garnish with coriander sprigs and lime wedges.

Chicken Salad

3 chicken breasts, boned and skinned

½ onion, chopped

1 carrot, chopped

150ml/¼ pint/⅔ cup chicken stock

few fresh herbs or a pinch of dried
mixed herbs

30ml/2 tbsp butter or sunflower
margarine

15ml/1 tbsp plain flour

30ml/2 tbsp frozen sweetcorn,
defrosted

3 slices bread

2 celery sticks

1 green dessert apple

1 red dessert apple

mixed green salad leaves and 1 small
tomato, cut into wedges, to serve

For the Dressing

45ml/3 tbsp natural yogurt

45ml/3 tbsp mayonnaise

5ml/1 tsp ground coriander

salt and freshly ground black pepper

1 Rinse the chicken breasts under
cold water and place in a
saucepan with the chopped onion
and carrot, stock and herbs. Cover
and cook for 15 minutes or until the
chicken is cooked through.

2 Cut one chicken breast into
small dice, cutting the other two
into larger pieces.

3 Strain the stock into a jug, finely
chop the carrot and remove and
discard the onion and herbs.

4 Preheat the oven to 190°C/
375°F/Gas 5. Melt 15ml/1 tbsp
of the butter or margarine in a small
saucepan, stir in the flour and then
gradually stir in the strained stock
and bring to the boil, stirring until
the sauce is thickened and smooth.
Add the finely diced chicken, carrot
and sweetcorn.

5 Cut the bread into three 7.5cm/
3in squares, cutting the
trimmings into tiny shapes using
cutters. Spread both sides of the
squares and one side of the tiny
bread shapes with the remaining
butter or margarine. Press the
squares into sections of a bun tray.

6 Bake the small shapes in the
oven for 5 minutes and the
squares for 10 minutes, until crisp
and golden brown all over.

7 Rinse and thinly slice the celery,
and quarter, core and chop half
of each of the apples.

8 To make the salad dressing,
blend the yogurt, mayonnaise,
ground coriander and salt and
pepper in a bowl. Add the thickly
diced chicken and the celery, pepper
and apples and mix together.

9 Tear the salad leaves into pieces
and arrange on the adults'
serving plates. Spoon the chicken
salad over the salad leaves.

10 Chop or process half of the
child's chicken mixture to the
desired consistency for the baby and
spoon into a dish with the tiny bread
shapes. Reheat the remaining
mixture if necessary, spoon into the
bread cases and arrange on a plate for
the toddler. Slice the two apple
halves, cutting the peel away from a
few slices and add to the children's
dishes. Check the temperature of hot
food before serving to children.

FISH DINNERS

Quick to make for everyday or a special occasion – here are some fish recipes for any day of the week. Choose from fresh, frozen or canned fish for added convenience. Forget about plain fishfingers and encourage the children to try homemade Fish Cakes with a tasty sauce for the adults, Paella, eye-catching Fishy Vol-au-Vents or colourful Salmon and Cod Kebabs.

Paella

400g/14oz cod fillet
115g/4oz fish cocktail or a mixture of prawns, mussels and squid
15ml/1 tbsp olive oil
1 onion, chopped
1 garlic clove, crushed
150g/5oz/⅔ cup long grain white rice
pinch of saffron or turmeric
few sprigs of fresh thyme or pinch dried
225g/8oz can tomatoes
½ red pepper, cored, seeded and chopped
½ green pepper, cored, seeded and chopped
50g/2oz frozen peas
30ml/2 tbsp fresh chopped parsley
salt and freshly ground black pepper

1 Remove any skin from the cod and place the fish cocktail in a sieve and rinse well with cold water.

2 Heat the oil in a frying pan, add the onion and fry until lightly browned, stirring occasionally. Add the garlic and 115g/4oz/½ cup of the rice and cook for 1 minute.

3 Add the saffron or turmeric, thyme, 2 of the canned tomatoes, 350ml/12fl oz/1½ cups water and salt and pepper. Bring to the boil and cook for 5 minutes.

4 Put the remaining rice and canned tomatoes in a small saucepan with 90ml/3fl oz/⅓ cup water. Cover the pan and cook for about 5 minutes.

5 Add 15ml/1 tbsp of the mixed peppers and 15ml/1 tbsp of the frozen peas to the small pan, adding all the remaining vegetables to the large pan. Place 115g/4oz of the fish in a metal sieve, cutting in half if necessary. Place above a small pan of boiling water, cover and steam for 5 minutes. Add the remaining fish to the paella in the large pan, cover the pan and cook for 5 minutes.

6 Add the fish cocktail to the paella, cover the pan and cook for a further 3 minutes.

7 Stir the chopped parsley into the paella and spoon on to two warmed serving plates for the adults. Spoon half of the tomato rice mixture and half the fish on to a plate for the child and process the remaining fish and rice to the desired consistency for baby. Spoon into a dish and check both children's meals for bones. Test the temperature of the children's food before serving.

Fish Cakes

450g/1lb potatoes, cut into pieces

450g/1lb cod fillet

75g/3oz prepared spinach leaves

75ml/5 tbsp full-fat milk

25g/1oz/2 tbsp butter

1 egg

50g/2oz/1 cup fresh breadcrumbs

25g/1oz drained sun-dried tomatoes

25g/1oz drained stuffed olives

115g/4oz Greek yogurt

3 tomatoes, cut into wedges

½ small onion, thinly sliced

15ml/1 tbsp frozen peas, cooked

sunflower oil, for frying

salt and pepper

lemon wedges and green salad, to serve

1 Half fill a large saucepan or steamer base with water, add the potatoes and bring to the boil. Place the cod in a steamer top or in a colander above the saucepan. Cover and cook for 8–10 minutes or until the fish flakes easily when pressed.

2 Take the fish out of the steamer and place on a plate. Add the spinach to the steamer, cook for 3 minutes until just wilted and transfer to a dish. Test the potatoes, cook for 1–2 minutes more if necessary, then drain and mash with 30ml/2 tbsp of the milk and the butter.

3 Peel away the skin from the fish, and break into small flakes, carefully removing any bones. Chop the spinach and add to the potato with the fish.

4 For baby, spoon 45ml/3 tbsp of the mixture into a bowl and mash with another 30ml/2 tbsp milk. Add a little salt and pepper, if liked, to the remaining fish mixture.

5 For the older child, shape three tablespoons of mixture into three small rounds with floured hands. For adults, shape the remaining mixture into four cakes.

TIP
Make sure you remove all bones from the fish.

6 Beat the egg and the remaining milk on a plate. Place the breadcrumbs on a second plate and dip both the toddler and the adult fish cakes first into the egg and then into the crumb mixture.

7 Chop the sun-dried tomatoes and stuffed olives and stir into the yogurt with a little salt and pepper. Spoon into a small dish.

8 Heat some oil in a frying pan and fry the small cakes for 2–3 minutes each side until browned. Drain well and arrange on a child's plate. Add tomato wedges for tails, peas for eyes.

9 Fry the adult fish cakes in more oil if necessary for 3–4 minutes each side until browned and heated through. Drain and serve with the dip, lemon and tomato wedges, thinly sliced onion and a green salad. Reheat the baby portion if necessary but test the temperature of the children's food before serving.

Fishy Vol-au-Vents

250g/9oz puff pastry, thawed if frozen

a little flour

beaten egg, to glaze

350g/12oz cod fillet

200ml/7fl oz/⅓ cup milk

½ leek

30ml/2 tbsp sunflower margarine

30ml/2 tbsp plain flour

50g/2oz fresh prawns

salt and pepper

broccoli, young carrots and mange tout, to serve

1 Preheat the oven to 220°C/ 425°F/Gas 7. Roll out the pastry on a lightly floured surface to make a rectangle 23 × 15cm/9 × 6in. Cut into three 7cm/3in wide strips.

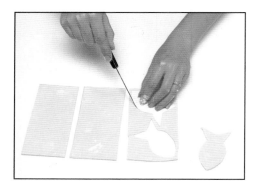

2 Cut two fish shapes from one strip. Knock up edges with a knife and cut an oval shape just in from the edge and almost through to the bottom of the pastry. Place the fish shape on a wetted baking tray.

TIP
You may find it easier to cut a fish shape out of paper and then use this as a template on the pastry.

3 Neaten the edges of the remaining pastry strips then cut smaller rectangles 1cm/½in in from the edge of both and remove.

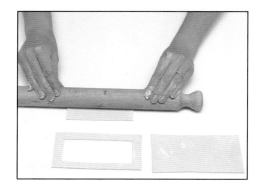

4 Roll out the smaller rectangles to the same size as the pastry frames. Brush the edges with a little egg and place the frames on top.

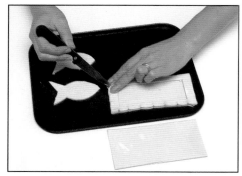

5 Transfer to the baking sheet. Knock up the edges of both rectangles with a small knife and flute between finger and thumb. Repeat the whole process with the third strip of pastry.

6 Stamp out tiny fish shapes from the pastry trimmings, rerolling if necessary, and place on the baking sheet. Brush the top of the vol-au-vents with beaten egg and cook for 10 minutes, remove the small fish and cook the larger vol-au-vents for a further 5 minutes until they are well risen and golden brown.

7 Meanwhile make the filling. Cut the fish in two and put in a saucepan with the milk. Halve the leek lengthways, wash thoroughly then slice and add to the pan.

8 Cover and simmer for 6–8 minutes or until the fish flakes easily when pressed with a knife. Lift out of the pan, peel away the skin and then break into pieces, removing and discarding any bones.

9 Strain the milk, reserving the leeks. Melt the margarine, stir in the flour and then gradually add in the milk and bring to the boil, stirring until thick and smooth.

10 Stir the fish into the sauce. Scoop out the centres of the fish-shaped vol-au-vents and fill with a little mixture. Spoon 30ml/2 tbsp of fish mixture into a baby bowl.

11 Add a prawn to each fish-shaped vol-au-vent and stir the rest into the saucepan with the leeks and a little seasoning. Heat gently and spoon into the two large pastry cases. Transfer to plates and serve with steamed vegetables.

12 Chop the baby portion with some vegetables and serve this with some tiny pastry fishes, if wished, in a small bowl. Check the temperature of the children's food before serving.

Salmon and Cod Kebabs

200g/7oz salmon steak

275g/10oz cod fillet

15ml/3 tsp lemon juice

10ml/2 tsp olive oil

2.5ml/½ tsp Dijon mustard

salt and freshly ground black pepper

350g/12oz new potatoes

200g/7oz frozen peas

50g/2oz/4 tbsp butter

15–30ml/1–2 tbsp milk

3 tomatoes, chopped, seeds discarded

¼ round lettuce, finely shredded

For the Mustard Sauce

1 sprig fresh dill

20ml/4 tsp mayonnaise

5ml/1 tsp Dijon mustard

5ml/1 tsp lemon juice

2.5ml/½ tsp dark brown sugar

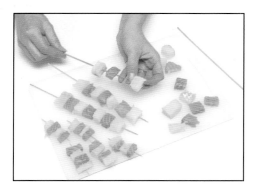

2 Cut a few pieces of fish into smaller pieces and thread on to five cocktail sticks. Thread the remaining fish pieces on to long wooden skewers.

3 Mix together the lemon juice, oil, Dijon mustard and a little salt and pepper to taste in a small bowl and set aside.

4 Finely chop the dill and place in a bowl with the other sauce ingredients and mix. Set aside.

5 Scrub the potatoes, halve any large ones and then cook in lightly salted water for 15 minutes or until tender. Place the peas and half of the butter in a frying pan, cover and cook very gently for 5 minutes.

6 Preheat the grill, place the kebabs on a baking sheet and brush the larger kebabs with the lemon, oil and mustard mixture. Grill for 5 minutes, until the fish is cooked, turning once.

7 Remove the fish from two cocktail sticks and mix in a small bowl with a few new potatoes and 15ml/1 tbsp of the peas. Chop or process with a little milk until the desired consistency is reached, then transfer to a baby bowl.

8 Arrange the toddler's kebabs on a small serving plate with a few potatoes, 30ml/2 tbsp peas and a small amount of chopped tomato. Test the temperature of the children's food before serving.

9 Add the tomatoes to the remaining peas and cook for 2 minutes, stir in the shredded lettuce and cook for 1 minute. Spoon on to serving plates for the adults, add the kebabs and potatoes and serve.

TIP
Depending on the age of the toddler, you may prefer to remove the cocktail sticks before serving.

1 Rinse the fish under cold water and pat dry. Cut the salmon steak in half, cutting around the central bone. Cut away the skin and then cut into chunky cubes making sure you remove any bones. Remove the skin from the cod and cut into similar sized pieces.

Tuna Florentine

50g/2oz pasta shapes

175g/6oz fresh spinach leaves

50g/2oz frozen mixed vegetables

200g/7oz can tuna fish in brine

2 eggs

buttered toast and grilled tomatoes, to serve

For the Sauce

25g/1oz/2 tbsp sunflower margarine

45ml/3 tbsp plain flour

300ml/½ pint/1¼ cups milk

50g/2 oz/½ cup grated Cheddar cheese

1 Cook the pasta in boiling water for 10 minutes. Meanwhile tear off any spinach stalks, wash the leaves in plenty of cold water and place in a steamer or colander.

2 Stir the frozen vegetables into the pasta as it is cooking, and then place the spinach in the steamer over the top. Cover and cook for the last 3 minutes or until the spinach has just wilted.

TIP
Frozen vegetables are usually more nutritious than fresh vegetables as they are frozen at their peak of perfection – they're a great time saver too.

3 Drain the pasta and vegetables through a sieve. Chop one quarter of the spinach and add to the pasta and vegetables, dividing the remainder between two shallow 300ml/½ pint/1¼ cup ovenproof dishes for the adults.

4 Drain the tuna and divide among the dishes and sieve.

5 Refill the pasta pan with water, bring to the boil then break the eggs into the water and gently simmer until the egg whites are set. Remove the eggs with a slotted spoon, trim the whites and shake off all the water. Arrange the eggs on top of the tuna for the adults.

6 To make the sauce, melt the margarine in a small saucepan, stir in the flour, then gradually add the milk and bring to the boil, stirring until thickened and smooth.

7 Stir the cheese into the sauce, reserving a little for the topping. Spoon the sauce over the adults' portions until the eggs are covered. Stir the children's pasta, vegetables and tuna into the remaining sauce and mix together.

8 Spoon the pasta on to a plate for the toddler and chop or process the remainder for baby, adding a little extra milk if necessary to make the desired consistency. Spoon into a dish. Test the temperature of the children's food before serving.

9 Preheat the grill, sprinkle the adults' portions with the reserved cheese and grill for 4–5 minutes until browned. Serve with buttered toast and grilled tomatoes.

VEGETABLE FEASTS

MORE AND MORE OF US ARE OPTING TO EAT LESS MEAT FOR EITHER HEALTH OR BUDGET REASONS. EATING LESS MEAT DOESN'T MEAN LOSING OUT ON FLAVOUR: FAR FROM IT! CHOOSE FROM MEALS SUCH AS CHEESE ON TOAST, CHEESY VEGETABLE CRUMBLE OR SLOW-COOKED SPICED VEGETABLE TAGINE.

Cheese on Toast

2 slices thick bread

2 slices thin bread

butter or sunflower margarine, for spreading

5ml/1 tsp Marmite

2 slices wafer thin ham

1 garlic clove, halved

15ml/1 tbsp olive oil

200g/7oz Cheddar cheese

1 tomato

6 stoned black olives

15ml/1 tbsp chopped fresh basil

a few red onion slices

cucumber sticks and green salad, to serve (optional)

2 Drizzle the oil over the thick slices of toast then rub with the cut surface of the garlic.

3 Thinly slice the cheese and place over all of the pieces of toast.

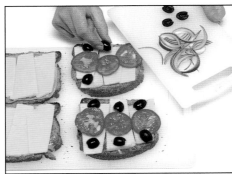

4 Slice the tomato and halve the olives. Arrange on the thick toast for the adults, season with pepper and add basil and onion.

5 Grill until the cheese is bubbly. Cut the plain cheese on toast into tiny squares, discarding crusts. Arrange on a small plate for baby and allow to cool. Cut the Marmite and cheese toast into triangles and arrange on a plate. Add cucumber sticks to the toddler's portion and test the temperature before serving.

6 Slice the adults' toast and arrange on plates with a green salad if liked, and serve.

1 Toast all the bread and spread the thin slices with butter or margarine. Spread one slice with Marmite and top this with ham.

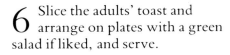

Penne with Tomato Sauce

150g/5oz penne or small pasta quills

1 onion

2 celery sticks

1 red pepper

15ml/1 tbsp olive oil

1 garlic clove, crushed

400g/14oz can tomatoes

2.5ml/½ tsp caster sugar

8 stoned black olives

10ml/2 tsp pesto

1.5ml/¼ tsp dried chilli seeds (optional)

50g/2oz/½ cup grated Cheddar cheese

salt and freshly ground black pepper

10ml/2 tsp freshly grated Parmesan cheese, to serve

green salad, to serve (optional)

2 Meanwhile chop the onion and the celery. Cut the pepper in half, then scoop out the core and seeds and dice the pepper finely.

3 Heat the oil in a second saucepan, add the vegetables and garlic and fry for 5 minutes, stirring until lightly browned.

6 Spoon 45–60ml/3–4 tbsp of the mixture into a bowl or processor and chop or process to the desired consistency for baby. Spoon 75–90ml/5–6 tbsp of the mixture into a bowl for the toddler.

1 Cook the pasta in salted boiling water for about 10–12 minutes until just tender.

4 Add the tomatoes and sugar and cook for 5 minutes, stirring occasionally until the tomatoes are broken up and pulpy.

5 Drain the pasta, return to the pan and stir in the tomato sauce.

7 Quarter the olives and stir into the remaining pasta with the pesto, chilli seeds if using, and a little salt and pepper. Spoon into dishes.

8 Sprinkle the grated Cheddar cheese over all the dishes, and serve the adults' portions with the Parmesan cheese and a green salad, if wished. Test the temperature of the children's food before serving.

Courgette Gougère

50g/2oz/4 tbsp butter

65g/2½oz plain flour

2 eggs

2.5ml/½ tsp Dijon mustard

115g/4oz Cheddar cheese

salt and pepper

For the Filling

15ml/1 tbsp olive oil

350g/12oz courgettes, sliced

1 small onion, chopped

115g/4oz button mushrooms, sliced

225g/8oz tomatoes, skinned and chopped

1 garlic clove, crushed

20ml/4 tsp chopped fresh basil

1 Preheat the oven to 220°C/ 425°F/Gas 7 and grease a large baking sheet. Place the butter in a medium-sized saucepan with 150ml/ ¼ pint/⅔ cup water. Heat gently until the butter has melted then bring to the boil.

2 Remove the pan from the heat and quickly add the flour, stir well then return to the heat and cook for 1–2 minutes, stirring constantly until the mixture forms a smooth ball. Leave to cool for 10 minutes.

3 Beat the eggs, mustard and a little salt and pepper together. Cut 25g/1oz of the cheese into small cubes and grate the rest.

4 Add 50g/2oz of the grated cheese to the dough and then gradually beat in the eggs to make a smooth, glossy paste.

5 Spoon the mixture into a large piping bag fitted with a medium plain tube. Pipe whirls close together on the baking sheet to make an 18cm/7in circular shape.

6 Pipe the remaining mixture into small balls for baby and into the older child's initials.

7 Sprinkle the top with the remaining grated cheese.

8 Bake for 15–18 minutes for the small shapes and 25 for the ring, until well risen and browned.

9 Heat the oil in a frying pan, add the courgettes and onion and fry until lightly browned. Add the mushrooms, fry for a further 2 minutes and then add the tomatoes. Cover and simmer for 5 minutes until the vegetables are cooked.

10 Spoon a little mixture into baby's bowl or into a processor and chop or process to the desired consistency. Serve with a few cheesy balls and a little diced cheese.

11 Spoon a little on to a plate for the toddler, add a cheesy initial and the remaining diced cheese. Test the temperature of the children's food before serving.

12 Transfer the cheesy ring to a serving plate and split in half horizontally. Stir the garlic, basil and a little extra seasoning into the courgette mixture and heat through. Spoon over the bottom half of the cheese ring and add top half. Serve the gougère immediately.

TIP
Beat the gougère mixture, cheese and eggs together in a food processor if preferred.

Cheesy Vegetable Crumble

1 onion

225g/8oz carrot

175g/6oz swede

175g/6oz parsnip

15ml/1 tbsp olive oil

220g/7½oz can red kidney beans

10ml/2 tsp paprika

5ml/1 tsp ground cumin

15ml/1 tbsp plain flour

300ml/½ pint/1¼ cups vegetable stock

225g/8oz broccoli, to serve

For the Topping

115g/4oz Cheddar cheese

50g/2oz wholemeal flour

50g/2oz plain flour

50g/2oz sunflower margarine

30ml/2 tbsp sesame seeds

25g/1oz blanched almonds

salt and freshly ground black pepper

1 Preheat the oven to 190°C/ 375°F/Gas 5. Peel and roughly chop the onion and peel and cut the carrot, swede and parsnip into small cubes. Heat the oil in a large pan and fry the vegetables for 5 minutes, stirring until lightly browned.

2 Drain the kidney beans and add to the pan with the spices and flour. Stir well, add the stock, then cover and simmer for 10 minutes.

3 Meanwhile make the topping. Cut a few squares of cheese for the baby and grate the rest. Put the two flours in a bowl, add the margarine and rub in with your fingertips until the mixture resembles fine breadcrumbs. Stir in the grated cheese and sesame seeds.

4 Spoon a little of the vegetable mixture into a 300ml/½ pint/ 1¼ cup ovenproof dish for the toddler. Spoon the remaining mixture into a 900ml/1½ pint/ 3¾ cup ovenproof pie dish for the adults, leaving a little vegetable mixture in the pan for the baby. Season the adults' portion with a little salt and pepper.

VARIATION
Lentil and Herb Crumble
Substitute canned lentils for the red kidney beans. Instead of the paprika and cumin, use 2 tbsp chopped, fresh mixed herbs. Add more fibre and nuttiness by replacing the plain flour with rolled oats.

5 Spoon 45ml/3 tbsp of crumble over the older child's portion. Roughly chop the almonds, add to the remaining crumble with a little salt and pepper and spoon over the large dish. Bake in the oven for 20 minutes for the small dish, 30 minutes for the large dish, until golden brown on top.

6 Add 90ml/3fl oz/⅓ cup water to baby's portion, cover and cook for 10 minutes, stirring occasionally, until the vegetables are very tender. Mash or process to the desired consistency and spoon into a bowl.

7 Cut the broccoli into florets and cook for 5 minutes or until tender; drain. Spoon the toddler's portion out of the dish and on to a small plate. Serve the broccoli to all members of the family, allowing baby to pick up and eat the broccoli and cubed cheese as finger food. Check the temperature of foods before serving to the children.

Note: Never give whole nuts to children under 5 as they may choke.

Vegetable Tagine

1 onion

225g/8oz carrots

225g/8oz swede

75g/3oz prunes

20ml/4 tsp olive oil

425g/15oz can chick-peas

2.5ml/½ tsp turmeric

10ml/2 tsp plain flour, plus extra for dusting

2 garlic cloves, finely chopped

450ml/¾ pint/1⅞ cups chicken stock

15ml/1 tbsp tomato purée

2cm/¾in piece fresh root ginger

2.5ml/½ tsp ground cinnamon

3 cloves

115g/4oz couscous

8 green beans

2 frozen peas

piece of tomato

knob of butter or sunflower margarine

salt and freshly ground black pepper

sprig fresh coriander, to garnish

1 Peel and chop the onion and peel and dice the carrots and swede. Cut the prunes into chunky pieces, discarding the stones.

2 Heat 15ml/3 tsp of the olive oil in a large saucepan, add the onion and fry until lightly browned. Stir in the carrots and swede and fry for 3 minutes, stirring.

3 Drain the chick-peas and stir into the pan with the turmeric, flour and garlic. Add 300ml/½ pint/1¼ cup of the stock, the tomato purée, and the prunes. Bring to the boil, cover and simmer for 20 minutes, stirring occasionally.

4 Place three heaped spoonfuls of mixture in a bowl or food processor, draining off most of the liquid. Mash or process and then form the mixture into a burger with floured hands.

5 Chop or process two heaped spoonfuls of mixture and sauce to the desired consistency for baby and spoon into a bowl.

6 Finely chop the root ginger and stir into the remaining vegetable mixture with the cinnamon, cloves, remaining stock and seasoning.

7 Place the couscous in a sieve, rinse with boiling water and fluff up the grains with a fork. Place the sieve above the vegetables, cover and steam for 5 minutes.

8 Fry the veggie burger in the remaining oil until browned on both sides. Trim and cook the beans and peas for 5 minutes. Drain and arrange on a plate like an octopus with a piece of tomato for a mouth and peas for eyes.

9 Stir the butter or margarine into the couscous and fluff up the grains with a fork. Spoon on to warmed serving plates for adults, add the vegetable mixture and garnish with a sprig of coriander. Check the temperature of the children's food before serving.

Stilton and Leek Tart

175g/6oz/1½ cups plain flour

75g/3oz/6 tbsp butter

1 carrot

2 slices ham

10cm/4in piece of cucumber

mixed green salad leaves, to serve

For the Filling

25g/1oz/2 tbsp butter

115g/4oz trimmed leek, thinly sliced

75g/3oz Stilton cheese, diced

40g/1½oz Cheddar cheese, grated

3 eggs

175ml/6fl oz/¾ cup milk

pinch of paprika

salt and pepper

1 Put the flour in a bowl with a pinch of salt. Cut the butter into pieces and rub into the flour with your fingertips until the mixture resembles fine breadcrumbs.

2 Mix to a smooth dough with 30–35ml/6–7 tsp water, knead lightly and roll out thinly on a floured surface. Use to line an 18cm/7in flan dish, trimming round the edge and reserving the trimmings.

3 Reroll the trimmings and then cut out six 7.5cm/3in circles using a fluted biscuit cutter. Press the pastry circles into sections of a bun tray and chill all of the tarts.

4 Preheat the oven to 190°C/375°F/Gas 5. For the filling melt the butter in a small frying pan and fry the leek for 4–5 minutes, until soft but not brown, stirring frequently. Turn into a bowl, stir in the diced Stilton, and then spread over the base of the large tart.

5 Beat the eggs and milk together in a small bowl and season with salt and pepper.

6 Divide the grated Cheddar among the small tartlet shells and pour some of the egg mixture over. Pour the remaining egg mixture over the leek and Stilton tart and sprinkle with paprika.

7 Cook the small tartlets for 15 minutes and the large tart for 30–35 minutes until well risen and browned. Leave to cool.

8 Peel and coarsely grate the carrot. Cut the ham into triangles for "sails". Cut the ham trimmings into small strips for the baby. Cut the cucumber into batons.

9 Place a spoonful of carrot, some cucumber, a small tart and some ham trimmings in the baby dish. Spread the remaining carrot onto a plate for the older child, place the tarts on top and secure the ham "sails" with cocktail sticks. Serve any remaining tarts next day. Cut the large tart into wedges and serve with a mixed leaf salad.

JUST DESSERTS

N O ONE CAN RESIST A DESSERT, ALTHOUGH LOOKING AFTER A YOUNG FAMILY MAY MEAN IT'S MORE OF A WEEKEND TREAT THAN AN EVERYDAY OCCURRENCE. SPOIL THE FAMILY WITH A SELECTION OF THESE TASTY HOT AND COLD PUDS FROM FRUITY EVE'S PUDDING AND PLUM CRUMBLE TO CHOCOLATE TRIFLE OR ORANGE AND STRAWBERRY SHORTCAKES – ALL GUARANTEED TO GET THEM CLAMOURING FOR SECONDS.

Bread and Butter Pudding

3 dried apricots

45ml/3 tbsp sultanas

30ml/2 tbsp sherry

7 slices white bread, crusts removed

25g/1oz/½ tbsp butter, softened

30ml/2 tbsp caster sugar

pinch of ground cinnamon

4 eggs

300ml/½ pint/1¼ cups milk

few drops of vanilla essence

pinch of cinnamon

pouring cream, to serve

2 Preheat the oven to 190°C/375°F/Gas 5. Spread the bread with butter and cut one slice into very small triangles. Layer in a 150ml/¼ pint/⅔ cup pie dish with plain apricots and sultanas and 5ml/1 tsp of the sugar.

4 Beat the eggs, milk and vanilla together and pour into the ramekin and two pie dishes.

1 Chop two dried apricots and place in a small bowl with 30ml/2 tbsp of the sultanas and the sherry. Set aside for about 2 hours. Chop the remaining apricot and mix with the remaining sultanas.

3 Cut the remaining bread into larger triangles and layer in a 900ml/1½ pint/3¾ cup pie dish with the sherried fruits and all but 5ml/1 tsp of the remaining sugar, sprinkling the top with cinnamon. Put the last 5ml/1 tsp sugar in a small ramekin dish.

5 Stand the ramekin dish in a large ovenproof dish and half fill with hot water. Cook the small pie dish and the ramekin for 25–30 minutes until the custard is just set, the larger pudding for 35 minutes until the bread is browned. Serve the adult portions with cream, if liked.

Eve's Pudding

500g/1¼lb cooking apples

50g/2oz/¼ cup caster sugar

50g/2oz frozen or canned
  blackberries

For the Topping

50g/2oz/4 tbsp butter or sunflower
  margarine

50g/2oz/¼ cup caster sugar

50g/2oz/⅓ cup self-raising flour

1 egg

½ lemon, rind only

15ml/1 tbsp lemon juice

icing sugar, for dusting

custard, to serve

1 Preheat the oven to 180°C/
350°F/Gas 4. Peel and slice the
cooking apples, discarding core, and
then place in a saucepan with the
caster sugar and 15ml/1 tbsp water.
Cover and cook gently for 5 minutes
until the apple slices are almost
tender but still whole.

2 Half fill a 150ml/¼ pint/⅔ cup
ovenproof ramekin with apple
for the toddler and mash 30ml/2 tbsp
apple for baby in a small bowl.

3 Put the remaining apple slices in
to a 600ml/1 pint/2½ cup
ovenproof dish. Sprinkle the
blackberries over the apple slices.

4 To make the topping, place the
butter or margarine, sugar, flour
and egg in a bowl and beat until
smooth. Spoon a little of the
pudding mixture over the toddler's
ramekin so that the mixture is
almost to the top of the dish.

5 Half fill three petits fours cases
with the pudding mixture.

6 Grate the lemon rind and stir
with the juice into the remaining
mixture. Spoon over the large dish,
levelling the surface.

7 Put the small cakes, toddler and
adult dishes on a baking sheet
and bake in the oven for 8–10
minutes for the small cakes, 20
minutes for the ramekin and 30
minutes for the larger dish, until
they are well risen and golden
brown.

8 Dust the toddler's and adults'
portions with icing sugar and
leave to cool slightly before serving
with the custard. Warm the baby's
portion if liked and test the
temperature before serving, with the
cakes taken out of their paper cases.

TIP
If you can't get blackberries then
use raspberries instead. There's no
need to defrost before using as
they will soon thaw when added
to the hot apple.

Orange and Strawberry Shortcakes

75g/3oz/6 tbsp plain flour

50g/2oz/4 tbsp butter

25g/1oz/2 tbsp caster sugar

grated rind of ½ orange

extra sugar, for sprinkling

For the Filling

175g/6oz Greek yogurt

15ml/1 tbsp icing sugar

250g/9oz strawberries

5ml/1 tsp Cointreau (optional)

2 sprigs of fresh mint

1 Preheat the oven to 180°C/ 350°F/Gas 4. Place the flour in a bowl, cut the butter into pieces, and rub into the flour with your fingertips until the mixture resembles fine breadcrumbs.

2 Stir in the sugar and orange rind and mix to a dough.

3 Knead the dough lightly then roll out on a floured surface to 5mm/¼in thickness. Stamp out four 9cm/3½in flower shapes or fluted rounds with a biscuit cutter and 12 small car, train or other fun shapes with novelty cutters, rerolling the dough as necessary.

4 Place on a baking sheet, prick with a fork and sprinkle with a little extra sugar. Bake in the oven for 10–12 minutes until pale golden, then leave to cool on the baking sheet.

5 For the filling, blend the yogurt with the sugar and wash and hull the strawberries. Pat dry. Reserve eight of the strawberries and process or liquidize the rest. Press through a sieve and discard the seeds.

6 For the adults, put 45ml/3 tbsp of the yogurt in a bowl and stir in the Cointreau, if using. Slice four strawberries and halve two, place on a plate and cover.

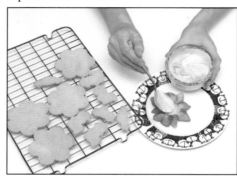

7 For the toddler, slice two of the strawberries and arrange in a ring on a small plate. Spoon 30ml/2 tbsp of yogurt into the centre of the ring and serve with three of the small biscuit shapes.

8 For baby, stir 15ml/1 tbsp of the strawberry purée into the remaining natural yogurt and spoon into a small dish. Serve with one or two of the small biscuit shapes.

9 To complete the adults' portions, spoon purée over two plates to cover completely.

10 Spoon the reserved yogurt over two biscuits, add the sliced strawberries and top with the other two biscuits. Arrange on plates and decorate with halved strawberries and tiny sprigs of mint.

TIP
If you find shortbread difficult to roll out then chill for 20 minutes. Knead lightly and roll out on a surface dusted with flour, dusting the rolling pin too.

Exotic Fruit Brulée

oil, for greasing

30ml/2 tbsp demerara sugar

1 ripe mango

1 kiwi fruit

1 passion fruit

30ml/2 tbsp icing sugar

350g/12oz Greek yogurt

1 Line a baking sheet with foil then draw round the tops of two 250ml/8fl oz/1 cup ramekin dishes and one 150ml/¼ pint/⅔ cup ramekin dish. Lightly brush with oil. Sprinkle the sugar inside each marked circle in an even layer.

2 Grill the sugar discs for 2–3 minutes or until it has melted and caramelized. Leave the discs to cool on the baking sheet.

3 Slice the mango either side of the central stone and then cut six thin slices for decoration, cutting away the skin. Cut the rest of the mango flesh from the stone, removing the skin, and finely chop one quarter, dividing between the baby bowl and small ramekin. Roughly chop the remainder and divide between the larger ramekins.

4 Peel the kiwi fruit, cut in half lengthways and then slice thinly. Reserve four half slices for decoration and divide the remaining fruit among the dishes, finely chopping the fruit for baby and toddler. Cut the passion fruit in half and using a teaspoon scoop and place the seeds in the adults' dishes.

5 Stir the icing sugar into the yogurt and mix 15ml/1 tbsp into the baby's portion. Spoon 30ml/2 tbsp yogurt over the toddler's dish and level the surface with a spoon.

6 Spoon the remaining yogurt into the other two ramekins, level the surface with a spoon and chill all of them until required.

7 When ready to serve, place the larger ramekins on two plates with the reserved mango and kiwi slices arranged around the sides. Peel the sugar discs off the foil and set on top of the adult and toddler portions. Serve the brulées immediately.

TIP
Traditionally the sugar topping on a brulée is made by sprinkling demerara sugar over the top of the dessert and then grilling, or by making a caramel syrup and then pouring it over the dessert. Making the topping on oiled foil is by far the easiest and most foolproof way. Add the cooled sugar discs at the very last minute so that they stay crisp and you get that wonderful mix of crunchy sugar, velvety yogurt and refreshing fruit.

Plum Crumble

450g/1lb ripe red plums

25g/1oz/2 tbsp caster sugar

For the Topping

115g/4oz/1 cup plain flour

50g/2oz/4 tbsp butter, cut into pieces

25g/1oz/2 tbsp caster sugar

10ml/2 tsp chocolate dots

75g/3oz marzipan

30ml/2 tbsp rolled oats

30ml/2 tbsp flaked almonds

custard, to serve

1 Preheat the oven to 190°C/ 375°F/Gas 5. Wash the plums, cut into quarters and remove the stones. Place in a saucepan with the sugar and 30ml/2 tbsp water, cover and simmer for 10 minutes.

2 Drain and spoon six plum quarters onto a chopping board and chop finely. Spoon the plums into a small baby dish with a little juice from the saucepan.

3 Drain and roughly chop six more of the plum quarters and place in a 150ml/¼ pint/⅔ cup ovenproof ramekin dish with a little of the juice from the saucepan.

4 Spoon the remaining plums into a 750ml/1¼ pint/3⅔ cup ovenproof dish for the adults.

5 Make the topping. Place the flour in a bowl, rub in the butter and then stir in the sugar.

6 Mix 45ml/3 tbsp of the crumble mixture with the chocolate dots then spoon over the ramekin.

7 Coarsely grate the marzipan and stir into the remaining crumble with the oats and almonds. Spoon over the adults' portion.

8 Place the toddler and adult portions on a baking sheet and cook for 20–25 minutes until golden brown. Leave to cool slightly before serving. Warm baby's portion if liked and check the temperature of the children's food before serving. Serve with custard.

TIP
Ovens can vary in temperature: with fan-assisted ovens you may need to cover the adults' dish with foil half way through cooking to prevent overbrowning.

Vary the fruits: cooking apples, peaches and pears also work well. If the plums are very sharp you may need to add a little extra sugar.

Chocolate and Orange Trifle

½ chocolate Swiss roll

3 clementines or satsumas

20ml/4 tsp sherry

50g/2oz dark chocolate

300ml/½ pint/1¼ cups ready-made custard

90ml/6 tbsp double cream

chocolate buttons or M & M's

3 Peel the remaining clementines, roughly chop and add to the adults' bowls, sprinkling a little sherry over each.

5 Spoon 30ml/2 tbsp of the custard into a small bowl for baby and arrange on the plate with the Swiss roll and clementines. Spoon a little more custard into the ramekin and then add the rest to the adult dishes, smoothing the surface.

1 Slice the Swiss roll, halve one slice and put on a plate for baby. Put a second slice into a ramekin dish for the older child and arrange the remaining slices in two dessert bowls for the adults.

2 Peel one clementine, separate into segments and put a few on the baby plate. Chop the rest and add to the ramekin.

4 Break the chocolate into pieces and melt in a bowl over a saucepan of hot water. Stir the melted chocolate into the custard.

6 Whip the cream until it just holds its shape. Add a spoonful to the ramekin dish and two or three spoonfuls each to the adults' dishes, decorating each dish with sweets. Chill the trifles until required.

TIP
Chocolate may be melted in the microwave in a microwave-safe bowl for 2 minutes on Full Power (100%), stirring thoroughly halfway through cooking.

Some shops sell ready-made chocolate custard, buy this if available and you're short of time.

Vary the fruit depending on what is in season, sliced strawberries and orange or sliced banana and fresh or canned cherries also work well.

Apple Strudel

450g/1lb cooking apples

50g/2oz dried apricots

45ml/3 tbsp sultanas

30ml/2 tbsp soft light brown sugar

30ml/2 tbsp ground almonds

5ml/1 tsp ground cinnamon

25g/1oz/2 tbsp butter

150g/5oz/3 sheets filo pastry, defrosted if frozen

icing sugar, to dust

pouring cream, to serve

1 Preheat the oven to 200°C/400°F/Gas 6. Peel, core and chop the apples and place 150g/5oz in a saucepan. Chop three apricots and add to the pan with 15ml/1 tbsp sultanas, 15ml/1 tbsp sugar and 15ml/1 tbsp water. Cover and simmer for 5 minutes.

2 Chop the remaining apricots and place in a bowl with the remaining apples, sultanas, sugar, ground almonds and cinnamon. Mix together well.

3 Melt the butter in a small saucepan or microwave in a microwave-proof bowl for 30 seconds on Full Power (100%).

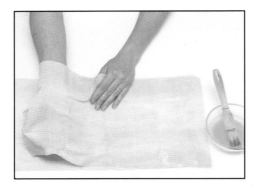

4 Carefully open out the pastry. Place one sheet on the work surface, brush with butter then cover with a second sheet of pastry and brush with more butter.

5 Spoon the uncooked apple mixture in a thick band along the centre of the pastry.

6 Fold the two short sides up and over the filling and brush with butter. Fold the long sides up and over the filling, opening out the pastry and folding the pastry for an attractive finish. Place on a baking sheet and brush with a little butter.

7 Brush a little of the butter over half of the third pastry sheet then fold the unbrushed half over the top to make a square. Brush again and cut into three equal strips.

8 Put a spoonful of apple at the base of each strip then fold the bottom right-hand corner up and over the filling to the left side of the strip to make a triangle.

9 Continue folding the pastry to make a triangular pasty. Repeat with the other strips.

10 Transfer to a baking sheet and brush with the remaining butter. Bake for 10 minutes, and the adults' strudel for 15 minutes, until golden and crisp. Dust with icing sugar and cool slightly.

11 Spoon the remaining cooked apple mixture into a baby dish. Mash the fruit if necessary. Transfer the toddler's portion to a plate. Slice the strudel thickly and serve with cream.

TIP
Filo pastry is usually sold frozen in 275g/10oz packs or larger. Defrost the whole pack, take out as much pastry as you need, then rewrap the rest and return to freezer.

Strawberry Pavlova

2 egg whites

115g/4oz caster sugar

2.5ml/½ tsp cornflour

2.5ml/½ tsp wine vinegar

For the Topping

150ml/¼ pint/⅔ cup double cream

250g/9oz strawberries

2 chocolate dots

green jelly sweet or little pieces
  of angelica

1 Preheat the oven to 150°C/
300°F/Gas 2 and line a baking
sheet with non-stick baking paper.

2 Whisk the egg whites until stiff
and then gradually whisk in the
sugar 5ml/1 tsp at a time. Continue
whisking until smooth and glossy.

3 Blend the cornflour and vinegar
and fold into the egg whites.

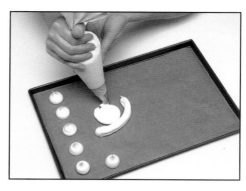

4 Spoon the mixture into a large
piping bag fitted with a
medium-sized plain tube. Pipe six
small dots for baby and a snail for
the older child with a shell about
6.5cm/2½in in diameter.

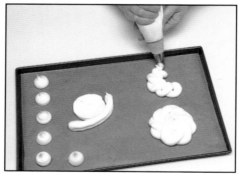

5 Pipe the remaining mixture into
two 10cm/4in swirly circles and
cook in the oven for 20–25 minutes
or until firm. Lift the meringues off
the paper and leave to cool.

TIP
For perfect meringues whisk the
egg whites in a dry greasefree
bowl. Remove any trace of yolk
with a piece of shell and whisk
until the peaks are stiff but still
moist looking. Add sugar gradually
and continue whisking until very
thick. Meringues may be made
1–2 days in advance. Cover with
greaseproof paper and decorate
just before serving.

6 Whip the cream until softly
peaking and spoon over the two
large meringues and the snail,
reserving about 15ml/1 tbsp for the
baby.

7 Rinse, hull and slice the
strawberries and arrange a few
in rings over the snail. Place the snail
on a plate and add chocolate eyes and
a slice from a green jelly sweet or a
little angelica for the mouth.

8 Add some strawberries to the
adult meringues then chop or
mash a few extra slices and stir them
into the reserved cream for the baby.
Spoon into a small dish and serve
with tiny meringues.

Poached Pears

3 ripe pears

450ml/¾ pint/1⅞ cup apple juice

5ml/1 tsp powdered gelatine

few drops of green food colouring

½ small orange, rind only

½ small lemon, rind only

10ml/2 tsp chopped glacé ginger

5ml/1 tsp clear honey

5ml/1 tsp cornflour

1 red liquorice bootlace

2 small currants

1 raisin or large currant

½ glacé cherry

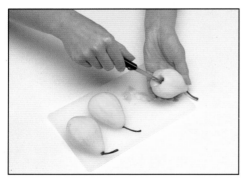

1 Peel the pears leaving the stalks in place. Cut a small circle out of the base of two pears and tunnel out the cores with a small knife.

2 Halve the remaining pear, discard the stalk, scoop out the core and then put them all in a pan with the apple juice. Bring to the boil, cover and simmer for 5 minutes, turning the pears once.

3 Lift out the pear halves and reserve, cooking the whole pears for a further 5 minutes or until tender. Remove with a slotted spoon and place on a serving dish reserving the cooking liquor. Finely chop one pear half and put into a small dish for baby. Reserve the other half pear for the toddler.

4 Pour half of the reserved apple juice into a bowl, place over a pan of simmering water, add the gelatine and stir until completely dissolved. Transfer to a measuring jug and pour half over the chopped pear and chill.

5 Stir a little green colouring into the remaining gelatine mixture, pour into a shallow dish or a plate with a rim for the toddler, and chill.

6 Cut away the rind from the orange and lemon and cut into thin strips. Add to the remaining apple juice with the ginger and honey and simmer over a gentle heat for 5 minutes until the rinds are soft.

7 Blend the cornflour with a little water to make a smooth paste, stir into the pan then bring the mixture to the boil and cook, stirring until thickened. Pour over the whole pears and leave to cool.

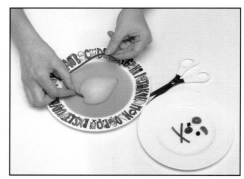

8 Place the toddler's pear, cut side down on the green jelly. To make a mouse, add a piece of red liquorice for a tail and two pieces for whiskers. Make two small cuts for eyes and tuck a currant into each slit. Add a raisin or large currant for the nose. Halve the cherry and use for ears. Serve slightly chilled with cream for the adults, if liked.

Peach Frangipane Tart

175g/6oz/1½ cups plain flour
75g/3oz/6 tbsp butter
For the Filling
50g/2oz/4 tbsp butter
50g/2oz/¼ cup caster sugar
1 egg
few drops of almond essence
50g/2oz/⅔ cup ground almonds
30ml/2 tbsp apricot jam
15ml/1 tbsp chopped candied peel
400g/14oz can peach slices in natural juice
15ml/1 tbsp flaked almonds
15ml/1 tbsp fromage frais or natural yogurt
icing sugar, for dusting
crème fraîche or natural yogurt, to serve

1 Place the flour in a bowl, cut the butter into small pieces and rub into the flour with your fingertips until the mixture resembles fine breadcrumbs.

2 Stir in 30–35ml/6–7 tsp water and mix to a smooth dough. Knead lightly on a floured surface and roll out thinly.

3 Lift the pastry over a rolling pin and use to line an 18cm/7in flan dish pressing down against the sides of the dish. Trim the top.

4 Reroll the trimmings and cut out twelve 5cm/2in circles with a fluted biscuit cutter, pressing into sections of a mini muffin tin. Chill the pastry for 15 minutes.

5 Preheat the oven to 190°C/ 375°F/Gas 5. Line the large flan dish with greaseproof paper and fill with baking beans. Cook for 10 minutes then remove the baking beans and paper and cook the shell for a further 5 minutes.

6 Meanwhile make the filling: cream together the butter and sugar in a bowl until light and fluffy. Beat the eggs and almond essence together then gradually beat into the creamed sugar. Stir in the ground almonds and set aside.

7 Divide the jam among the small and large tarts, spread it into a thin layer in the large tart and sprinkle with candied peel. Spoon all of the almond mixture over the top and level the surface.

8 Drain the peaches and set aside three slices for baby. Cut three slices into chunky pieces and divide among six of the small tarts, leaving six with just jam.

9 Arrange the remaining peaches over the top of the large tart and sprinkle with flaked almonds.

10 Cook the small tarts for 10 minutes and the large tart for 25 minutes until the filling is set and browned. Dust the large tart with icing sugar and leave to cool slightly.

11 Purée the reserved peaches for baby in a processor or liquidizer. Mix with fromage frais or yogurt and spoon into a small dish. Serve with two plain jam tarts.

12 Arrange a few fruity jam tarts on a plate for the toddler. Cut the large tart into wedges and serve with crème fraîche or natural yogurt for the adults. Test the temperature of the children's tarts before serving.

Honey and Summer Fruit Mousse

10ml/2 tsp powdered gelatine

500g/1¼lb bag frozen summer
   fruits, defrosted

20ml/4 tsp caster sugar

500g/1¼lb tub fromage frais or
   Greek yogurt

150ml/¼ pint/⅔ cup whipping
   cream

25ml/5 tsp clear honey

1 Put 30ml/2 tbsp cold water in a cup and sprinkle the gelatine over making sure that all grains of gelatine have been absorbed. Soak for 5 minutes, then heat in a saucepan of simmering water until the gelatine dissolves and the liquid is clear. Cool slightly.

2 For baby, process or liquidize 50g/2oz of fruit to a purée and stir in 5ml/1 tsp sugar. Mix 45ml/3 tbsp fromage frais or yogurt with 5ml/1 tsp sugar in a separate bowl.

3 Put alternate spoonfuls of fromage frais or yogurt and purée into a small dish for baby. Swirl the mixtures together with a teaspoon. Chill until required.

4 For the toddler, process 50g/2oz fruit with 5ml/1 tsp sugar. Mix 60ml/4 tbsp fromage frais or yogurt with 5ml/1 tsp sugar then stir in 10ml/2 tsp fruit purée.

5 Stir 5ml/1 tsp gelatine into the fruit purée and 5ml/1 tsp into the fromage frais or yogurt mixture. Spoon the fruit mixture into the base of a perspex beaker or glass and chill until set.

6 For the adults, whip the cream until softly peaking. Fold in the remaining fromage frais or yogurt and honey and add the remaining gelatine. Pour into two, 250ml/8fl oz/1 cup moulds. Chill until set.

7 Spoon the remaining fromage frais mixture over the set fruit layer in the serving glass for the toddler and chill until set.

8 To serve, dip one of the dishes for the adults in hot water, count to 15, then loosen the edges with your fingertips, invert on to a large plate and, holding mould and plate together, jerk to release the mousse and remove mould. Repeat with the other mould. Spoon the remaining fruits and some juice around the desserts and serve.

TIP
If you have a small novelty mould you may prefer to set the toddler's portion in this rather than a glass or plastic container.

Make sure the gelatine isn't too hot before adding to fromage frais or yogurt or it may curdle.

Hot Fruit Salad with Ginger Cream

2 oranges

4 bananas

50g/2oz stoned dates

30ml/2 tbsp sultanas

25g/1oz/2 tbsp butter

45ml/3 tbsp demerara sugar

30ml/2 tbsp Cointreau or brandy

vanilla ice cream, to serve

For the Ginger Cream

150ml/¼ pint/⅔ cup double cream

15ml/1 tbsp chopped glacé ginger

grated rind of 1 orange

3 Peel and slice the bananas. Roughly chop the dates. Arrange a few orange segments, slices of banana, pieces of date and a few sultanas on a plate for the baby.

5 Stir the Cointreau or brandy into the pan, bring to the boil then quickly light with a match. Spoon the fruit into a serving bowl as soon as the flames subside. Serve the adults' flambéed fruit with ginger cream and the toddler's portion with ice cream. Test the temperature of toddler's portion before serving.

1 First make the ginger cream: whip the cream until it just holds its shape, then finely chop the ginger and stir into the cream with the orange rind. Spoon into a small serving dish.

4 Melt the butter in a frying pan, add the remaining fruit and fry for 2 minutes, then stir in the sugar and cook for 2 minutes, until lightly browned. Spoon a few pieces of fruit into a bowl for the toddler.

TIP
Flaming dishes look so spectacular but are really very easy to do. The secret is to add the spirit to hot liquid, bring the liquid back to the boil then quickly light it with a match and stand back. The flames will subside in a minute or two. But if you are at all worried they can be quickly extinguished by covering the pan with a lid or baking sheet.

2 Cut a slice off the top and bottom of each orange and cut the remaining peel and pith away in slices. Cut into segments.

INDEX

ACKNOWLEDGEMENTS

- The Department of Health
- National Dairy Council Nutrition Service
- Dr Nigel Dickie from Heinz Baby Foods
- The British Dietetic Association
- The Health Visitors' Association
- Broadstone Communications for their invaluable help supplying the Kenwood equipment for recipe testing and photography
- Hand-painted china plates, bowls and mugs from Cosmo Place Studio, 11 Cosmo Place, London WC1N 3AP. Tel. 0171 278 3374
- Tupperware for plain-coloured plastic bowls, plates, feeder beakers and cups
- Cole and Mason for non-breakable children's ware
- Royal Doulton for Bunnykins china
- Spode for blue and white Edwardian Childhood china

PICTURE CREDITS

The publishers would like to thank the following
for additional images used in the book:

Key: t = top; b = bottom; l = left; r = right.

Bubbles: pages 8 b, 11 t (Lois Joy Thurston); page 9 t (Nikki Gibbs);
pages 14 t, 13 t (Ian West); 15 br (Jacqui Farrow).
Lupe Cunha: page 8 t.
Reflections/Jennie Woodcock: pages 15 br, 9 b, 10 bl br.
Timothy Woodcock: page 11.

MODELS

The publishers would like to thank the following children and
adults for being such wonderful models: Maurice Bishop,
Andrew Brown, Penny and Chloe Brown, Daisy May Bryant,
April Cain, Helen and Matthew Coates, Cameron Gillis,
Jamie Grant, Sandra and George Hadfield, Ted Howard,
Emily Johnson, Huw and Rhees Jones, Key, Stephen, Charlie
and Genevive Riddle, William Lewis, Sadé Walsh, Lionel and
Lucy Watson, Lily May Whitfield, Philippa Wish, James Wyatt.